THEORY OF THE EARTH;

OR, THE

PERIODICALLY RECURRING SUPERFICIAL CHANGES, OR GEOLOGICAL REVOLUTIONS, IN THE EARTH'S CRUST;

ALSO,

THE CHANGES IN THE ORGANIC WORLD,

INDICATED IN THE GEOLOGICAL RECORD;

TOGETHER WITH THE PROXIMATE CAUSE OF THE SAME, VIZ: THE CLIMATAL VICISSITUDES OF FORMER TIMES, CONSIDERED WITH REFERENCE TO THE PROPER MOTION OF THE EARTH, INVOLVED IN THE ASTRONOMICAL APPEARANCE KNOWN AS THE "DIMINUTION OF THE OBLIQUITY OF THE ECLIPTIC TO THE EQUATOR."

BY

JAMES BRADFORD BABBITT.

"Whatever is real is rational."

"We must look to existing causes for the explanation of past geological events."

TAUNTON, MASS.:
AUTHOR'S EDITION.
1875.

Electrotyped at the Boston Stereotype Foundry,
19 Spring Lane.

PREFACE.

THE peculiar circumstances under which the following sketch of what appears to be the true terrestrial system has been written, emboldens me to ask in its behalf for some mitigation of the received rules of literary criticism. No ordinary contingency could have induced me to assume the role of authorship. Unaccustomed to the use of the pen, and unable fully to command the uninterrupted leisure requisite to the proper performance of a work of this nature, the propriety of my making the attempt at first appeared questionable. The obvious importance of the subject, however, seemed to demand its publication, and being naturally unwilling to delegate the task to another hand, the only alternative was to contribute my best endeavors, and, whatever the result, rely on the generous consideration of those interested in discussions of this kind. The work is now completed, and if the effort shall prove so far successful as to render the general tenor of the arguments clearly intelligible, it is all I can reasonably hope to have accomplished.

In this essay, which at best is but an incomplete outline of the new system, I have judged it proper to attempt the exploration of but little ground; and but few books have been consulted. Sir Charles Lyell's " Principles of

Geology " and the late Professor Louis Agassiz's " Journey in Brazil " are the works chiefly relied upon for the facts which underlie our hypothesis, and free use has been made of their contents. Herschel, Mitchel, Vose, and Hind have been referred to in the discussion of the astronomical points involved; and I will here add that it is the results of Charles Darwin's investigations concerning the origin of species, and those of Professor Agassiz's observations of the phenomena that remain of the latest glacial period in tropical South America, together with the perusal of Vose's chapter on the " Diminution of the Obliquity of the Ecliptic," to which I am chiefly indebted for the inception of the within theory.

CONTENTS.

CHAPTER V.

A BRIEF REVIEW OF VARIOUS HYPOTHESES ADVANCED TO ACCOUNT FOR THE CLIMATAL VICISSITUDES OF FORMER TIMES.

CHAPTER VI.

ON THE PROPER AND DETERMINATE CAUSE OF PRIMARY CLIMATAL CHANGES.

CHAPTER VII.

GENERAL SUMMARY AND CONCLUSION.

INTRODUCTION.

It seems quite certain, judging from past experience, that every new truth, on its first appearance, must encounter more or less opposition. What men believe — what they have accepted as truth, even when the only basis for such belief consists in the fact that it has been generally taught and received as truth — often receives the sanction of faith in equal measure with the most exact and certain knowledge. It is an exceedingly difficult task, no doubt, to view with equanimity the overthrow of cherished theories and opinions, and accord a candid and impartial hearing to those which seek to occupy their place. It has often happened that when the world has been, as it were, expecting the advent of some new doctrine or discovery, the reality, when it made its appearance, was so different from the fanciful image conjured up by the imagination, that the mass of mankind, even in the face of the most positive evidence, has seemed almost involuntarily to shrink from even its contemplation.

We are not far removed from the times when to advocate views materially opposed to those generally prevailing, especially if such views happened to be antagonistic to popular prejudice, would be very likely to subject the offender to quite serious consequences. The spirit of the present age, as we know full well, would be far from making it a felony, were the world ignorant of the fact, to proclaim the discovery of the diurnal revolution of the earth; still, if such theory should seem to be in opposition to what we conceive to be the established order of nature, and contrary also to the teachings of both our spiritual and scientific guides, might not the announcement be received with a contemptuous hostility, or an indifferent neglect, quite as galling as any amount of personal indignity? But however much of this spirit may still exist and be liable to develop itself in such cases, the liberal sentiments which adorn the works of the greatest intellects of the century are a sufficient assurance that, in the present age, any earnest and honest endeavor to extend the bounds of human knowledge, will somewhere meet with attention and encouragement, so far, at least, as the effort may tend to the elucidation of the great problems of nature. Who can read the following passage from Herbert Spencer, and continue to hesitate as to the expediency of giving full expression to his highest convictions of truth, not only in questions relating to social philosophy, but those also connected with the physical sciences?

"Whoever hesitates to utter that which he thinks the highest truth, lest it should be too much in advance of the time, may reassure himself by looking upon his acts from an impersonal point of view. Let him duly recognize the fact that opinion is the agency through which character adapts external arrangements to itself, — that his opinion rightly forms part of this agency, — is a unit of force, constituting with other such units the general power which works out social changes, — and he will perceive that he may properly give full utterance to his inmost convictions, leaving it to produce what effect it may. It is not for nothing that he has in him these sympathies with some principles and repugnance to others. He, with all his capacities, and aspirations, and beliefs, is not an accident, but a product of the time. He must remember that, while he is a descendant of the past, he is a parent of the future, and that his thoughts are as children born to him, which he may not carelessly let die. He, like every other man, may properly consider himself as one of the myriad agencies through whom works the Unknown Cause; and when the Unknown Cause produces in him a certain belief, he is thereby authorized to profess and act out that belief. For, to render in their highest sense the words of the poet, —

'Nature is made better by no mean
But nature makes that mean: over that art
Which you say adds to nature, is an art
That nature makes.'

"Not as adventitious, therefore, will the wise man regard the faith which is in him. The highest truth he sees he will fearlessly utter; knowing that, let what may come of it, he is thus playing his right part in the world."

In reference to the general unwillingness to receive new truths, Sir Charles Lyell says, "We are sometimes tempted to ask whether the time will ever arrive when science shall have obtained such an ascendency in the education of the millions that it will be possible to welcome new truths, instead of always looking upon them with fear and disquiet, and to hail every important victory gained over error, instead of resisting the new discovery, long after the evidence in its favor is conclusive. The motion of our planet round the sun, the shape of the earth, the existence of the antipodes, the vast antiquity of our globe, the distinct assemblages of species of animals and plants by which it was successively inhabited, and, lastly, the antiquity and barbarism of Primeval Man, — all these generalizations, when first announced, have been a source of anxiety and unhappiness. The future, now opening before us, begins already to reveal new doctrines, if possible, more than ever out of harmony with cherished associations of thought. It is, therefore, desirable when we contrast ourselves with the rude and superstitious savages who preceded us, to remember, as cultivators of science, that the high comparative place which we have reached in the scale of being has been gained, step by step, by a conscientious study of natural phenomena, and by fearlessly teaching the doctrines to which they point. It is by faithfully weighing evidence without regard to preconceived notions, by earnestly and patiently searching for what is true, not what we wish to be true, that we have attained that dignity which we may in vain hope to claim through the rank of an ideal parentage."

Baron Von Humboldt, the author of "Cosmos," recognizes the fact that any sincere effort of the human intellect to extend the bounds of knowledge is entitled to respectful consideration, as follows: —

"It is unsuitable to the spirit of the age to regard with distrust any attempted generalization of views, or investigation in the paths of reasoning or induction. Nor is it consonant with a due estimate of the dignity of the human intellect, and the relative importance of the faculties with which we are endowed, to condemn at one time severe reason applied to the investigation of causes and their connection, and at another that exercise of the imagination which is often precursive to discoveries — for the achievement of which the imaginative power is an essential auxiliary."

These extracts, a few of many at hand, indicate the existence of a sentiment highly favorable to scientific progress among those who, through their works, occupy the most advanced and exalted positions in the modern field of philosophical research. As a sacred duty, he who has reason to believe that the results of his study and meditations are calculated to advance the cause of truth is called upon to give fearless and earnest utterance to his thoughts. That they will be accorded a consideration commensurable with the promise they may give of important results, and that popular prejudice, or the authority of opinion, will not be allowed to work their detriment, can scarcely be a matter of doubt.

The interests of knowledge — the development of truth — is, above all, the aim of paramount importance. Neither what we have believed, nor what others have believed to be true, nor what we or others wish to be true, but what is true, is the ultimate goal to which our investigations are to tend. The efforts of the master minds of the age to remove all obstacles whatever to the progress of science, are already giving indications of their efficiency. "Science emancipated from the trammels of prejudice and authority," in the words of Professor Huxley, "is slowly bringing about the greatest intellectual revolution the world has yet seen. She is teaching the world that the ultimate court of appeal is observation and experiment, and not authority; she is teaching it to estimate the value of evidence; she is creating a firm and living faith in the existence of immutable moral and physical laws, perfect obedience to which is the highest possible aim of an intelligent being."

The novel, and I may venture to add, unexpected character of the conclusion to which our present inquiries lead, may fairly excuse a certain amount of surprise and incredulity in the public mind upon its first announcement. My own experience, the first effect produced upon my own mind by the idea, has prepared me to expect as much.

Unquestionably, the bare statement that our planet has a motion, determining in stated times complete revolutions transverse to its diurnal rotation, unaccompanied with the facts and processes leading thereto, or, indeed, until the applicability and pertinency of such facts and processes come to be fairly and fully appreciated, will meet with an emphatic and decided negative. Rash — absurd — impossible — are the epithets to be expected under the circumstances. But as further contemplation brings to view the ready and natural solution the system affords to the gravest and most perplexing scientific problems, and glimpses are attained of the vast generalizations it authorizes, and the ease with which all geological phenomena may be made conformable to it, and reduceable to a rational general system, it does seem as if the hasty verdict must be set aside. The novelty of the new system, it will be remembered, does not necessarily imply error; neither does its antagonism to received theories and opinions in any degree qualify or affect its truth; and it will be seen that in a matter of such scientific importance, first impressions are not to have weight.

The claims of a system giving fair promise of such far-reaching results are not to be lightly thrust aside. Even if the evidence adduced in this volume, upon an impartial examination, shall be found inadequate to warrant a favorable verdict, before a final adverse judgment is recorded, a full and exhaustive re-survey of the whole ground must be made. But there appears little danger of being reduced to this extremity. If the fundamental principles or laws of nature, and the inherent properties of matter, are stable and unchanging, the transverse rotation of the earth is as certain, as fixed a fact, as that the solar presence and absence constitute the terrestrial day and night, or that the direct and indirect action of the sun's rays determines, respectively, summer and winter.

While the whole mass of evidence derived from natural phenomena appears arranged on the affirmative side of the proposition, in opposition we have only the authority of mere opinion. Men who, by means of signal services rendered, have attained deserved eminence in the field of scientific investigation, became somehow impressed with the belief that the present inclination of the earth's axis is essential to what they have termed the "stability of nature," and have argued to that end. We hold, however, that opinions and impressions, no matter by whom they may be or have been entertained, are not of themselves to have weight in inquiries like that in which we now propose to engage.

The attributes of exactitude and certainty characterizing positive knowledge are sometimes accorded to speculative and empirical notions, leading to the acceptance of such notions as undoubted truth. Errors of this nature, sanctioned, perhaps, by long acquiescence, and based generally upon the opinion of some person justly distinguished in the annals of science, have in the past often proved the most difficult of removal of all the obstacles in the way of scientific progress. It is, indeed, not surprising that the immediate disciples of a great pioneer in the realms of knowledge, familiar with his discoveries, and intelligently following the infallible methods through which they were attained, should be led frequently to attach undue importance to mere expressions of his opinion. "Who," they may be supposed to exclaim, "shall presume to dispute the dictum of our intellectual giant? What lesser light shall dare point to spot or blemish on the sun?" Hence comes the authority of opinion; hence the large areas of ignorance, the impenetrable wastes of doubt and perplexity which mar the broad fields of human knowledge. He who above all the men of his time, from the extent and variety of his knowledge of the material world, may be considered most competent to the attainment of a comprehensive and just view of the whole domain of physical science, in "Cosmos," gives expression to a consciousness of this unsatisfactory state of affairs.

"It remains to be seen," he says, "whether we can hope by the operation of thought to reduce the immense diversity of phenomena comprehended by the cosmos to a unity of principle similar to that presented by the evidence of what are specially called 'rational truths.' In the present state of our empirical knowledge, at least, we dare not entertain such a hope. Experimental sciences founded on observation of the external world cannot aspire to completeness; the nature of things and the imperfection of our organs are alike opposed to it. We shall never succeed in exhausting the inexhaustible riches of nature, and no generation of men will ever be able to boast of having comprehended all phenomena." And again: "We are yet very far from the time, even supposing it possible that it should ever arrive, when a reasonable hope could be entertained of reducing all that is perceived by the senses to the unity of a single principle. The complication of the problem and the immensurable extent of the cosmos seem to forbid the expectation of such success in the field of natural philosophy being ever achieved by man; but," he continues, "the partial solution of the problem — the tendency towards a general comprehension of the phenomena of the universe — does not the less continue to be the high and enduring aim of all natural investigation."

As long as the authority of opinion shall be held paramount to rational deductions from observed facts, or while it shall be allowed to prevent such deductions from being made, "the complication of the problem" will continue to exist, and the "tendency towards a general comprehension of the phenomena of the universe" will be retarded. If, on the other hand, we ignore all doubtful data, and, confining ourselves strictly to known facts, conduct our inquiries in accordance with the principles laid down by Descartes in his "Discourse touching the Method of using Reason rightly, and of seeking Scientific Truth," we cannot fail to keep on solid ground. The central propositions of the whole discourse, in the language of Huxley, are these: "There is a path that leads to truth so surely that any one who will follow it must needs reach the goal, whether his capacity be great or small. And there is one guiding rule by which a man may always find this path, and keep himself from straying when he has found it. This golden rule is, Give unqualified assent to no propositions but those the truth of which is so clear and distinct that they cannot be doubted."

In harmony with this golden rule, we shall endeavor to present, as a basis for our argument, "propositions so clear and distinct that they cannot be doubted." Between these propositions, their corollaries, and well-ascertained facts, constituting the positive evidence in the case, and corroborative testimony of more or less probability bearing thereon, the line of demarcation will, if possible, be kept constantly in view.

In her present condition, Science may be represented as looking and hoping for the promulgation of the within theory. Having penetrated far beyond the point at which the discovery should, naturally, have been made, each step in advance is now taken with anxiety and hesitancy. Like a general operating far within the enemy's country, upon an inadequate and unprotected base, her attention and efforts, which should be concentrated upon the front, are distracted, and in a measure neutralized by apprehensions of disaster in the rear, and the whole movement seems thereby to be involved in fear and uncertainty. If she is expected to continue her march to new victories over ignorance and error, her basis of operations must be further extended and fortified, her lines of communication cleared and rendered secure. Every strategic point in the advance should be occupied; and if an invulnerable position, the key to and commanding the whole line, shall be brought to the notice of those intrusted with her interests, neither a fancied impracticability, nor the nature of the instrumentality pointing to it, can justify them in refusing or neglecting to secure it.

Biologists, says Huxley, are at present generally agreed that "the manifold varieties of animal and vegetable forms have not either come into existence by chance, nor result from capricious exertion of creative power; but that they have taken place in a definite order, the statement of which order is what men of science term a natural law." While expressing the profoundest ignorance of the laws determining the origin of species, they demonstrate their existence by their specific effects, as Leverrier demonstrated the existence of the planet Neptune in advance of its actual discovery.

The Duke of Argyle, according to Lyell, has observed that "we know nothing of the natural forces by which new forms of life are called into being. But he admits that the introduction of new species, to take the place of those that have passed away, is a work which has been not only so often, but so continuously repeated, that it does suggest the idea of having been brought about through the instrumentality of some natural process."

Huxley, in his vigorous style, adverts to the probable discovery of this process, or law: "Whether such a law is to be regarded as an expression of the mode of operation of natural forces, or whether it is simply a statement of the manner in which a supernatural power has thought fit to act, is a secondary question so long as the existence of the law and the possibility of its discovery by the human intellect are granted. But he must be a half-hearted philosopher who, believing in that possibility, and having watched the gigantic strides of the biological sciences during the last twenty years, doubts that science will, sooner or later, make this further step, so as to become possessed of the law of evolution of organic forms — of the unvarying order of that great chain of causes and effects of which all organic forms, ancient and modern, are the links." I cannot refrain from adding the concluding sentence of the paragraph. "And then, if ever," he says, "we shall be able to discuss with profit the questions respecting the commencement of life, and the nature of the successive populations of the globe, which so many seem to think are already answered."

And now, to the discovery of this great law — "the law of evolution of organic forms" — I make claim. I believe that the within theory furnishes the key not only to "that great chain of cause and effect of which all organic forms are the links," but also that it furnishes full, accurate, and rational solution to the obscure and perplexing problems relating to the inorganic world and the ever-varying conditions by which, in times past, it has been surrounded. The contrast between the simple and natural processes of our system, and the partial, far-fetched, and unsatisfactory methods of solution, which, against the dictates of their better judgment, so many have felt constrained to accept, must, as we proceed, be evident to every candid mind.

The full relative importance of the discovery, time alone can completely determine; and before concluding these introductory remarks with a statement of the line of argument intended to be pursued, I desire to say that it is with full reliance on the intelligence and liberal spirit of the age in which we live that I submit my generalization of natural phenomena to the public scrutiny.

Evidence will be introduced, in the first place, to show

that the same areas of the earth's surface have been successively subjected to widely different climatal conditions during the various periods of the world's past history, and that those differences are of so radical and pronounced a character that they properly may be denominated primary effects, and as such cannot be logically referred to secondary causes.

Although, from the axiomatic character of the proposition, evidently a work of supererogation, an effort will be made to show that the sun is an unchanging body, and that the amount of light and heat derived from it by the earth is constant, and the effect, under like conditions, always the same. Such being the case, it will be held that a definite, invariable result must follow the presence of the sun, and the direct action of the solar rays upon the earth's surface, such result being an abundant development of vegetable and animal life; while, on the contrary, the absence of the sun, or the indirect action of his rays, must as unavoidably be followed by processes of glaciation and dearth of life.

It will be shown that both these classes of phenomena are to be met with in all parts of the earth: evidence abounds tending to show the prevalence of former summers in polar regions, where now perpetual winter reigns, and proofs quite as conclusive, indicate that periods marked by comparatively severe winters have been experienced in the tropics, where now perpetual summer is to be met with; while in intermediate latitudes, the traces of ancient vicissitudes of temperature, far exceeding anything now known, are everywhere present, and have long been recognized and commented on by geologists.

A review of the efforts that have been made to explain these appearances, under the assumption that the earth and sun have always held, approximately, their present relative positions, will, it is believed, show that they, as well as all others that may hereafter be attempted under that hypothesis, are, and of necessity must be, irrational, impossible, and absurd.

The consistent and reasonable view will be advanced that wherever the fossil remains of plants and animals requiring a large amount of the solar influence for their development are found, that region, in their epoch, must have enjoyed, during some portion of the year, the direct action of the sun's rays, that is, they must have been vertical, or nearly so. Therefore, if certain geological formations occurring in polar regions shall be found to contain remains of the above description, it will be held as a legitimate conclusion therefrom that when they were living and growing beings, the earth must have assumed such a position with regard to the sun as to admit of those organisms receiving the usual supply of light and heat required for their development. On the other hand, wherever unquestionable indications of ice action, or glaciation, shall be found, such indications will be taken as proof either of the entire absence of the sun during some part of the year, or, if not below the horizon, that a large inclination of his rays must have been the cause of such phenomena. If, therefore, geological indication of extensive glacial action shall be found to exist at ordinary levels within the torrid zone, it will be held as competent evidence that at the era to which such indications may be referred, the position of the earth with regard to the sun was such that, during a portion of the year, the sun must have been absent or his rays must have exhibited a large inclination to the plane of the horizon within those areas where those indications occur.

It will be further shown the terrestrial changes of position indicated by the above-adduced facts are due to an almost inconceivably slow rotatory motion of the earth proceeding at a right angle to the axis of its diurnal revolution; a motion so slow that millions of years are required to complete a single revolution. The apparent movement of the ecliptic, resulting from the real motion of the earth in question, known to astronomers as the "diminution of the obliquity of the ecliptic," continued indefinitely, is the only one possible, consistent with the plan of the solar system by which these changes can have been produced.

It is assumed that, as we proceed, the fact will become developed, that geological periods, practically as well as theoretically, are definite, statedly recurring, equal lapses of time; each one of such periods representing the amount required to include a complete cycle of climatal changes, such as result from the transverse revolution of the earth; and it will also be seen that the several layers or series of strata, which in the earth's crust constitute a grand geological group, are the direct consequences of one of these cycles of climatal vicissitudes.

An effort will be made, as far as the means now at hand will permit, to show that all the climatal conditions consequent upon this movement throughout each period of change, instead of being, in a general sense, detrimental to organic life, by continually inducing, or, rather, necessitating, variations of habit and structure to meet the ever-varying conditions of life, have constituted the grand controlling cause by which low primitive organic types have been modified and improved to those of the present age; and that, contemporaneously with these modifications in its living inhabitants, the physical world, by the same means, and in the same progressive manner, has been adapted to the sustenance and fitted for the habitation of the various tribes by whom it has been successively occupied.

THEORY OF THE EARTH.

CHAPTER I.

EVIDENCE SHOWING THE PREVALENCE OF DIVERSE CLIMATAL CONDITIONS IN FORMER TIMES. — SIR CHARLES LYELL'S RETROSPECT.

EMPIRICAL AND ABSOLUTE KNOWLEDGE CONTRASTED IN A GENERAL VIEW OF NATURAL PHENOMENA. — DEFINITION OF THE TERM CLIMATE, AS USED IN THIS WORK. — PRIMARY CLIMATAL VARIATION NOW PERCEPTIBLE ONLY IN POLAR LOCALITIES. — RATE AT WHICH PRIMARY CLIMATIC CHANGE PROCEEDS. — AGE OF STONE AND ITS SUBDIVISIONS : THE NEOLITHIC AGE. — THE REINDEER PERIOD AND THE PALÆOLITHIC AGE. — CLIMATE OF THE TIMES OF THE MAMMOTH. — CAPACITY OF ANIMALS TO ADAPT THEMSELVES TO EXTREME VARIATIONS OF CLIMATE. — SIBERIAN FOSSILS.

It is natural for man, reasoning from the experience acquired during the brief portion of time allotted to him as the term of his existence, to conclude that the material world, from its first creation to the present time, has presented substantially the same appearance that it now does, and that the multifarious conditions on which depend the existence and well-being of the myriad living forms inhabiting it, throughout all past time, must have remained nearly, if not identically, the same. From youth to age, within the scope of his threescore and ten years, he has beheld no change in the general aspect of Nature. The physical conformation of the globe has remained the same. The same continents and oceans, islands and seas, mountains, plains, and rivers, that diversified its surface in his youth, retain their places in his old age. The sun, moon, and stars pursue the same courses, in a firmament presenting continually the same appearance. Day and night, summer and winter, continue to succeed one another with undeviating regularity; and the same forms of vegetable and animal life continue to inhabit and perpetuate their kind within those regions of the earth to which they are specifically adapted.

An intelligent being, reasoning from these appearances, might suppose himself justified in the conclusion that the operations of nature are of a fixed and stable character, admitting of no variation whatever. When to the fruits of his own experience he adds the concurrent testimony of his immediate ancestors, his conviction must be materially strengthened; and still further, when on consulting the written records of mankind from their very beginnings, he should find therein, as he would, much to support, and nothing to invalidate it, he might well consider the theory of the stability of nature satisfactorily and firmly established. What he throughout his whole life has seen, what his ancestors before him have seen, in fact, what all mankind have seen, must, indeed, be true.

As he progresses in knowledge, however, he soon learns to regard with distrust evidence resting entirely on the operation of the senses. By means of the invention and use of infallible instruments, he learns with surprise, that his visual organs, for instance, which he had supposed so true and reliable, have not always conveyed to his mind correct impressions of either the position or magnitude of objects. The delicate sense of touch, with which he is endowed, and on which he has placed so much reliance, is also found to be no sure guide; for the infallible thermometer shows him that a body maintaining the same degree of temperature may, under different circumstances, seem either hot or cold to him, as the case may be. By these and other like facts, he learns at last to distinguish between empirical or apparent truth and that which is exact or absolute; and he comes now, naturally enough, to doubt the truth of his former conclusions, founded, as they were, entirely on the evidence of sense.

Let him now be supposed to study geologically the crust of the earth, and his doubts must soon give way to certainty. Instead of the supposed stability of nature, he discovers a condition of incessant mobility, of continual change. He finds that the present configuration or superficial aspect of the globe, instead of having prevailed from the beginning, is, on the contrary, a comparatively modern affair — a single view in the grand cosmographical panorama, differing alike from each and all of many others that have gone before, and others which are to follow it. He discovers that in former remote periods of the earth's history, oceans and seas have rolled over the continents and islands he, his ancestors, and all mankind had previously supposed contemporaneous with its creation. He learns of the former existence of ranges of lofty mountains that have crumbled to dust, and have been levelled with the plain, long before those he now beholds began to raise their summits above the clouds. He discovers that ancient rivers have flowed over old continents, differing entirely in shape and superficial contour from the present land masses, whose entire courses have been obliterated, and whose existence is now known only by the vast quantities of sedimentary material brought down and deposited by them at their mouths.

In addition to the knowledge thus acquired of these successive changes in the arrangement of the matter forming the earth's crust, he finds from the inspection of fossilized remains there imbedded, that the living beings, or forms by which the world was formerly inhabited, are not identical with those now seen; and his former conclusion that the present inhabitants of the earth are the unchanged lineal descendants of individuals originally created identically the same as themselves, must, therefore, be abandoned as erroneous. Furthermore, careful physiological analyses of these remains, and just comparisons founded on peculiarities of structure common to them and to their nearest living representatives, taken in connection with the locality, north or south, in which they are found, together with other collateral testimony, bring to his knowledge the additional fact that, in climatal conditions also, the various epochs of the past present as marked a contrast with the present era as he had before found to subsist in either of the former instances.

The result, then, is the contemplation, in each of the various intervals of the earth's past history, of a world differing from the present alike in superficial structure, surrounding conditions, and inhabitants. The material — the primal elements of which it is composed — remains the same, but, kaleidoscope-like, it is continually assuming new forms, and entering into new combinations.

No change in the material world from any given status to another can transpire except through the intervention of adequate cause. Premising, therefore, that ordinary meteorological agents, acting with the various degrees of force possible under possible variations of temperature, are sufficient to produce, with exception of those depending on subsidence and upheaval, the before-mentioned superficial changes in the earth's crust, and also that the changing conditions of life induced by such possible variations of temperature are the primary cause of modification in the organic world; it is proposed, first, to consider the reality of former vicissitudes of climate, their nature and effect, and endeavor to trace, if possible, a chronological synchrony, or correspondence in time, between them and the progress of the terrestrial motion to which we believe they are properly referable. Our chief reliance, in questions of fact, will be upon the results of Sir C. Lyell's labors, as given in the tenth and eleventh chapters of "Principles of Geology," in which the subject of former climatal changes is fully discussed.

It may here be stated that the term *climate*, as used in this work, is, in general, to be understood as referring more particularly to states and changes of temperature, and mean temperature, than to those other elements included in its more extended sense.

It can scarcely be affirmed that any noticeable changes of climate, except those slight modifications which may be referred to certain local influences, have been detected in those parts of the earth inhabited by civilized man within the historical period. On the contrary, it may with truth be said that no such changes have taken place. According to Mosaic chronology, the grape and olive have flourished in Palestine for three thousand years; and from known peculiarities of those plants relative to the amount of heat required for their successful culture, the conclusion has been reached that the mean annual temperature cannot appreciably have varied there during that time. But if we turn from the warm to the polar regions of the earth, which obviously would first show the effect of diminishing obliquity, and where slight changes of temperature, especially a reduction of the same, must largely affect the relations of the organic world, we discover facts tending to show that climate is there undergoing a change, the rate of which is sufficiently rapid to materially affect, in the course of a few generations, the welfare of the races of men by whom those regions are inhabited.

In the account of his memorable expedition to the polar seas in search of Sir John Franklin, Dr. Kane has informed us that the igloës, or stone huts of the native Esquimaux, are to be found in locations far north of any now inhabited by them, and that their traditions indicate, with comparative certainty, that their by no means remote ancestors had been driven, by the increased rigor of the climate, from the more northern places to those now occupied; and he became convinced that the same cause must, in a short time, necessitate either a farther migration southward, or bring about their complete extermination.

"There is no doubt on my mind," says Dr. Kane,* "that at a time within historical and even recent limits, the climate of this region" (the shores of Smith's Sound, latitude 78° N.) "was milder than it is now. I might base this opinion on the fact, abundantly developed by our expedition, of a secular elevation of the coast line; but, independently of the ancient beaches and terraces and other geological marks which show that the shore has risen, the stone huts of the natives are found scattered along the line of the bay in spots now so fenced in by ice as to preclude all possibility of the hunt, and of course of habitation by men who rely on it for subsistence.

"Tradition points to these as once favorite hunting-grounds near open water. At Rensselaer Harbor, called by the natives Aunatok, or the Thawing Place, we met with huts in quite tolerable preservation, with the stone pedestals still standing, which used to sustain the carcasses of captured seals and walrus. Sunny Gorge, and a large indentation in Dallas Bay which bears the Esquimaux name of the Inhabited Place, showed us the remains of a village, surrounded by the bones of seals, walrus, and whales — all now cased in ice. In impressive connection with the same facts, showing not only the former exten-

* Arctic Explorations, vol. i. p. 308.

sion of the Esquimaux race to the higher north, but the climatic changes which may perhaps be in progress there, is a sledge runner which Mr. Morton saw on the shores of Morris Bay, in latitude 81°. It was made of the bone of a whale, and worked out with skilful labor."

On page 158 of the second volume occurs a further description of the Dallas Bay remains. "On the south-east corner of this bay, where some low islands at the mouth of the fiord formed a sort of protection against the north wind, was a group of Esquimaux remains — huts, cairns, and graves. Though evidently long deserted, my drivers" (native Esquimaux) "seemed to know all about them, for they suspended the hunt around the bergs to take a look at these evidences of a by-gone generation of their fathers.

"There were five huts with two stone pedestals for the protection of meat, and one of those strange little kennels which serve as dormitories when the igloë is crowded. The graves were higher up the fiord: from these I obtained a knife of bone, but no indications of iron.

"These huts stood high up upon a set of shingle terraces similar to those of Rensselaer Bay. The ice-belt at their foot was old and undisturbed, and must have been so for years; so too was the heavy ice of the bay. Yet around these old homesteads were bones of the seal and walrus, and the vertebræ of a whale, similar to that at the igloë of Aunatok. There must have been both open water and a hunting-ground around them, and the huts had in former days been close upon the water-line. 'Una suna nuna,' 'What land is this, Kalutuna?' I did not understand his answer, which was long and emphatic; but I found from our interpreter that the place was still called 'the inhabited spot;' and that a story was well preserved among them of a time when families were sustained beside its open water, and musk-ox inhabited the hills."

The bearing of the foregoing facts is plain. Assuming the annual amount of heat received by the earth from the sun to be always the same, a decrease in inclination of the terrestrial axis must have the effect to accumulate heat in the direction of the equator, at the expense of polar areas. But from the small proportion of the earth's surface comprised within the limits of the polar circles, as compared with that embraced within the torrid and temperate zones, amounts of heat which, taken from the former, would there produce sensible results, when distributed over the whole extent of the latter, cannot be supposed capable of producing any appreciable effect. The surface of the earth, also, being less convex near the poles than elsewhere, owing to its oblato-spheroidal figure, as the terrestrial axis of rotation approaches perpendicularity with the ecliptic, the value of the mean inclination of the sun's rays from the vertical must increase in an accelerating ratio, as compared with such an increase in lower latitudes. Thus the effect of diminishing inclination, accelerated as above described, may be assumed as sufficient to produce, within a century or two, sensible climatic changes in polar latitudes, while the correlative variation in tropical and temperate limits would not become observable in periods short of tens of thousands of years at least.

Before entering upon our contemplated survey of former diversities of climate, we ought to accustom our minds to the consideration of the rate at which the determining motion or cause progresses. It is necessary that we should acquire as accurate an idea as possible of its almost infinitesimal slowness.

The centennial amount of the diminution of the obliquity of the ecliptic was, at first, stated by astronomers to be 51″; but later and more reliable calculations have reduced the value of it to 48″; that is, the earth's axis becomes by 48″, in each one hundred years, nearer to perpendicularity with the ecliptic; and in reverse order, each one hundred years in our contemplated retrospect is to be considered as adding 48″ to the present amount of obliquity, viz., 23° 28′. If we divide the number of seconds contained in a degree by 48, we find that it requires seventy-five centuries, or seven thousand five hundred years, to change the direction of the earth's axis one degree. Now, when we consider the slight importance, as affecting climate, attributed by Lyell to the admitted amount of this motion, — he, on the authority of Sir John Herschel, assuming it to be a mere oscillation of the pole, extending, possibly, for three or even four degrees about a common mean, and intimating that a variation of at least four degrees would be required, conjointly with various other supposititious causes to make it of geological interest,* — when we come to view it alone as the primary and only cause, wholly unaided by, and independent of, all other agencies, we are enabled to realize, in some measure, the almost inconceivably slow rate at which its effect on the terrestrial climate must proceed. It may safely be assumed, therefore, that while the effect of 5°, equal in time to 37,500 years, might be barely perceptible in middle latitudes, 10°, or 75,000 years, at least, would be required for it to bring about marked changes. The present value of the inclination being about 23° 28′, by the above process we find that a period of 176,000 years will be required to bring the ecliptic coincident with the equator, and, retrospectively, another of 499,000 years must have elapsed since the coincidence of the polar axis with the same — intervals of time so vast that the human mind is entirely incapable of forming any adequate conception of their duration.

Granted the inherent susceptibility of organic forms to modification, in however slight a degree, what changes are not possible in such stupendous epochs?

Finding ourselves now in possession of a rule by which we may, approximately, locate in time any era of the past, the conditions of which shall enable us to determine with any degree of accuracy the direction of the earth's axis, or, rather, the relative positions of the earth and sun, we proceed with our investigations.

We have seen that, excepting in extreme polar areas, no fundamental changes of climate have taken place upon the earth within the historical period; and, in accordance with our mode of computation, we may conclude that none would be observable were the time embraced therein quadrupled. We observe, therefore, in the times immediately anterior to the historical, according to Lyell,† "no indications of any marked divergence from the present condition of things, whether in the memorials of the age of bronze or in those of the neolithic age which preceded it," — the term "Neolithic" being used, as the author observes in an appended note, "for this more modern age of stone, calling the older stone period, that in which

* Prin. of Geol., vol. i. p. 294. † Prin. of Geol., vol. i. p. 174.

man was contemporary with many extinct mammalia, 'Palæolithic.'"

The next antecedent era, included in and intermediate between the above-mentioned subdivisions of the age of stone, "is that designated by the late M. Lartet 'the Reindeer period,' when that northern animal, together with several others fitted for a cold climate, extended its range to the foot of the Pyrenees. The mammoth and cave lion, quadrupeds more characteristic of an anterior period, have been found sparingly in this fauna, and another extinct quadruped, the Irish elk, or gigantic deer."

Up to the present time, geologists have been unable to discover data of any kind by which the amounts of time embraced within the various geological periods could be determined, and mere conjecture has been their only resource. "Whenever any thoughtful geologist is asked," says Professor Huxley,* "what may be the approximate value in time of a 'great epoch'—whether it means a hundred years, or a thousand, or a million, or ten million years,—his reply is, 'I cannot tell;'" and the same observation has been applicable to the lesser divisions of time we are now considering. Although our system enables us, with accuracy, to locate in time certain marked eras, such, for instance, as the intermediate point of the glacial period, in regard to these subdivisions of so-called post pliocene times, we can only form judgments of more or less accuracy, according to the amount of information we may possess in relation to them. Evidently, there is at present less danger of overrating than underrating the time; and, finding in the reindeer period, in the extended southerly range of northen types of animals, the first marked effect of increased obliquity in middle latitudes, we may conclude that such increase must have been somewhere near 10°, equal in time to 75,000 years; and this may be taken as an approximation to the true date of the reindeer period.

It is to be remembered that we are not, as has been supposed, approaching, in the glacial epoch, a period characterized by an absolute decrease in the mean temperature of the whole earth. On the contrary, we are to

* Lay Sermons, Addresses, and Reviews, p. 209.

regard that epoch as one of highly contrasted seasons, in which exceedingly warm summers were succeeded by winters of arctic severity. Without entering into any discussion of the probable climatal effect of an increase of 20° in the annual course of the sun, north and south, its ascent to a height of more than 33° above the horizon, during the summer, at the poles, makes it appear quite possible that in the reindeer period, the progenitors of the present Esquimaux tribes were able to subsist within nearly if not all the regions of the polar circles.

The determination of solar heat from the poles in the direction of the equator, consequent on an increased annual mean inclination of the sun's rays caused by the gradually decreasing inclination of the earth's axis of diurnal rotation, is shown by the migration southward, in comparatively modern times, of the marine molluscous fauna of northern coasts. In Norway there are ancient sea-beaches which are now elevated six and seven hundred feet above the level of the ocean, "in which," says Lyell, "the shells are identical with those now living, although their geographical distribution has somewhat altered, the fossil species constituting an assemblage which at present characterizes the sea several degrees farther north." * On the supposition that the average rise of the land has been at the rate of two and a half feet in a hundred years, a period of 28,000 years has been required to raise some of these beaches from the sea level to their present altitude, and effect a change of climate, as above indicated, equal to the difference existing between places separated by several degrees of latitude. At the present known rate of progression, namely, 48″ per century, 28,000 years would change the direction of the terrestrial axis about 3° 6′, which might possibly suffice in those latitudes to accomplish this amount of climatal change; although, when we regard the changing conditions that have caused this southerly migration as depending less on an absolute decrease in mean temperature than on the colder and colder winters of an increasing obliquity, this lapse of time may seem too short. By referring to chapter xxxi. vol. ii. of the "Principles of Geology,"

* Prin. of Geol., vol. i. p. 133.

where the rising of land in Scandinavia is treated of at length, the reader will learn that the above estimated rise of two and a half feet per century is probably much too large, it being quite possible the same may not have been more than one fifth as much. The time, then, when those beaches constituted the actual coast line of Norway, instead of being 28,000 years ago, may have been anywhere from that number up to 140,000. Assuming, as a mean, that the actual rise has been one half of the largest estimate, or one and one quarter feet per century, and the time, in consequence, 56,000 years, the increase in the degree of inclination of the terrestrial axis would be somewhat in excess of $7\frac{1}{2}°$ over the present amount. We may with safety consider even so slight a change as this implies in the relative positions of the earth and sun, as sufficient to produce the observed effect; first, by the increased heat of the summers, and secondly, by the tendency of the movement to produce a more equal distribution of solar heat throughout all latitudes. There need be no hesitation, therefore, in accepting the fact of such change in climate as legitimate evidence of the persistive or rotatory nature of this motion of the earth. As furnishing us with the first indication in temperate latitudes of an approach towards the peculiar conditions of the glacial period, we may assign as the time when those ancient beaches formed the sea-coast line of the country, a point in the geological record somewhat intermediate between the neolithic and the earlier division of the age of stone.

Passing over this intermediate portion of the age of stone, "which," says Lyell, "is as yet but vaguely and imperfectly defined, we come to the older stone age, or 'Palæolithic period,' comprising the ancient river-gravels of Amiens and Abbeville in France, and of Salisbury and Bedford in England, and the superficial deposits of many other parts of Europe. Here, for the first time in our retrospect, we encounter the bones of a large number of extinct species of the genera elephant, rhinoceros, bear, tiger, and hyena, associated with the remains of living animals and of man. The human relics consist almost entirely in North-western Europe, of unpolished flint implements of a type different from those of the later or neolithic era, implying a less advanced state of civilization. The gravels containing such works of art and bones of extinct animals belong to a time when some of the minor features of the physical geography were different from those now characterizing the same part of Europe, a discordance which does not hold true of the more modern or neolithic times. The valleys of the more ancient of the two periods had not acquired their present width, depth, and outline. The bones of man and rude works of art occur also in caves associated with the remains of mammalia similar to those of the palæolithic gravels above mentioned. The enormous volume of alluvial matter formed in the channels of the old rivers, the contorted stratification of some parts of such alluvium, and the large size of many of the transported stones which it contains, imply a climate which generated much snow and ice in winter, and a mean annual temperature lower than that now found in the same parts of Europe." *

The above-cited facts are clearly indicative of a progressive change of climate, such as must ensue from the continued operation of the cause we have assigned. We observe in the palæolithic age abundant remains of numerous species of the larger mammals, animals that were, evidently, especially adapted to the widely diversified seasons of the glacial period, which was now approaching its termination; for we see them gradually decrease in number as the contrasts of seasons diminish through this and subsequent periods, until they finally become extinct. The periodical congregation, in caverns, of large numbers of many species of animals, is illustrated by the cave deposits of this era; and we behold, too, the rude men of the time seeking, in natural and artificial subterranean habitations, refuge both from the rigorous cold of the winters, which generated much ice and snow, and the excessive heats of the midsummers, intensified as they were by the increased altitude and long continuance of the sun above the horizon. The sudden liquefaction, on the approach of summer, of the large deposits of snow and ice generated during the winter, by enormously increasing the volume of water in the rivers, may be considered as furnishing a cause amply sufficient, even

* Prin. of Geol., vol. i. p. 175.

if the annual amount of effect was comparatively small, operating as it did throughout so vast an interval of time, to account for the superficial changes mentioned, due to fluviatile action; such as the deposition of large beds of alluvial matter, and the effects of the process of denudation by which the physical features of the valleys suffered so much change. The world at present furnishes no instance by which we may correctly estimate the value of these agencies in effecting the superficial terrestrial transformations of the glacial period. While it is true that certain localities now experience winters much more severe than those of the places mentioned in Palæolithic times, the fact that they are not followed, as were the latter, by more or less sudden transitions to the heat of tropical summers, is sufficient to determine the absence of this class of phenomena at the present time.

Another remarkable confirmation of the correctness of our views relative to the climatal peculiarities of the earlier portion of the age of Stone, occurs on page 568, vol. ii., of "Principles of Geology." In alluding to the drift of the south of Hampshire, in which flint implements of the Palæolithic age have been found, in order to explain the destruction of large masses of chalk, and the spreading of the flinty material originally dispersed in layers through it, over the ancient surface, enveloping at the same time the implements before mentioned, we must have recourse, he says, to ice action. "An extreme climate, causing a vast accumulation of snow during a cold winter, and great annual floods when this snow was suddenly melted in the beginning of the warm season, may best account," he thinks, for the phenomena.

This transfer of material by diluvial agency, a process, when proceeding on an extensive scale, we may suppose to be peculiar to glacial times, indicates that we have, in the earlier part of the age of Stone, already reached the immediate confines of the so-called Post Pliocene Glacial period; and it is this agency, gradually increasing in efficiency as the tens of thousands of years slowly roll by, that, although producing but little change in any one year or century perhaps, in the grand aggregate of more than half a million years, has completely revolutionized the superficial aspect of the globe.

If we assign as the probable date of the earlier part of the age of Stone that obtained by doubling the one assumed for the later, Palæolithic times go back one hundred and fifty thousand years from the present; and from thence to the intermediate point of the Ice period remains the immense interval of three hundred and forty thousand years; and this latter number, vast as it is, must be doubled in order to arrive at the full duration of the drift or glacial interval of the great Pliocene period.

From the identical nature of the phenomena characterizing this glacial interval, geologists have been unable to draw any very definite lines of division within it. The climate of the supposed later portions of it is considered by Lyell under the heads of "Climate of the Mammoth and its Associates," and "Climate of European Drift and Cave Deposits." It scarcely need be stated that these terms designate no well-defined periods of time. The animals referred to, the mammoth and its contemporaries, are, we believe, to be looked upon as the modified descendants of those living in the ages preceding the drift, when climate was essentially the same as now; the species that have become extinct being those whose constitution and habits became by degrees so closely adapted to the peculiar conditions of the glacial epoch, that on the gradual modification of those conditions, consequent on the return of the former climatic state, they became exterminated; while the species that have continued in existence to the present time are those which were less closely adapted to such conditions, or were possessed of greater plasticity of constitution.

"Geologists," says Lyell, "when they first examined the fossils of the drift, approached the subject with the fullest conviction on their minds that the climate of the globe in the olden times was warmer than it is now. This opinion they had legitimately derived from the study of the Tertiary and Secondary rocks, and when they encountered the bones of the elephant, rhinoceros, hippopotamus, lion, tiger, and hyena plentifully entombed in the old river gravels above mentioned, and in the contemporaneous mud and breccia of caverns, they concluded, without hesitation, that as all the genera alluded to are now characteristic of warmer latitudes, their presence was in perfect harmony with the received doctrine. The fact that the numerous land and fresh-water shells accompanying the same fossils were almost without exception identical with those now inhabiting the same country, ought doubtless to have served as a warning against the belief in a hotter climate; but the well-known forms of many large and conspicuous mammalia made a greater impression on their minds than the comparatively diminutive mollusca, with which few were familiar. The late Dr. Fleming, however, before the notion had gained ground that a glacial epoch had intervened between tertiary and historical times, called in question, in 1829, the opinion that the bones of the elephant and rhinoceros, and other associated pachyderms and beasts of prey, implied a tropical climate. A near resemblance, he observed, in form and osteological structure is not always followed in the existing mammiferous fauna by a similarity of geographical distribution; and we must therefore be on our guard against deciding too confidently, from mere analogy of anatomical structure, respecting the habits and physiological peculiarities of *species* now no more. 'The zebra,' he remarked, 'delights to roam over the tropical plains; while the horse can maintain its existence throughout an Iceland winter. The buffalo, like the zebra, prefers a high temperature, and cannot thrive even where the common ox prospers. The musk-ox, on the other hand, though nearly resembling the buffalo, prefers the stinted herbage of the arctic regions, and is able, by its periodical migrations, to outlive a northern winter. The jackal (*Canis aureus*) inhabits Africa, the warmer parts of Asia, and Greece; while the isatis, or arctic fox (*Canis lagopus*), resides in the arctic regions. The African hare and the polar hare have their geographical distribution expressed in their trivial names;' and different species of bears thrive in tropical, temperate, and arctic latitudes.

"Other writers soon followed up the same line of argument, and Mr. Hodgson, among others, in his account of the mammalia of Nepal, stated that the tiger was sometimes found at the very edge of perpetual snow in the Himalaya. Pennant had previously mentioned, that it had been seen among the snows of Mount Ararat in Armenia, and later authorities have placed it beyond all doubt that a species of tiger identical with that of Bengal is common in the neighborhood of Lake Aral, near Sussac, in the forty-fifth degree of North latitude. Humboldt remarks, that the part of Southern Asia now inhabited by this Indian species of tiger is separated from the Himalaya by two great chains of mountains, each covered with perpetual snow, — the chain of Kuenlen, lat. 35° N., and that of Mouztagh, lat. 42°, — so that it is impossible that these animals should merely have made excursions from India, so as to have penetrated in summer to the forty-eighth and fifty-third degrees of North latitude. They must remain all the winter north of the Mouztagh, or Celestial Mountains. The last tiger killed, in 1828, on the Lena, in lat. 52¼°, was in a climate colder than that of St. Petersburg and Stockholm.

"A species of panther (*Felis irbis*), covered with long hair, has been discovered in Siberia, evidently inhabiting, like the tiger, a region north of the Celestial Mountains, which are in lat. 42°.

"In regard to the climate of the living elephant, the Rev Robert Everest observes, that the greatest elevation at which it is found in a wild state is in the north-west Himalaya, at a place called Nahun, about 4,000 feet above the level of the sea, and in the 31st degree of N. lat., where the mean yearly temperature may be about 64° Fahrenheit, and the difference between winter and summer very great, equal to about 36° F., the mouth of January averaging 45°, and June, the hottest month, 91° F.

"Von Schrenck, writing in 1858, announced that in Amoorland, part of North-Eastern Asia, then recently annexed to the Russian Empire, no less than 34 out of 58 living quadrupeds are identical with European species. Among those which are not European, some are arctic, others of tropical forms; in illustration of which, he states that the Bengal tiger, ranging sometimes northwards as far as lat. 42°, subsists chiefly on the flesh of the reindeer, while on the other hand, the small tailless hare or pika occasionally wanders from its polar haunts to parts of Amoorland as far south as 48°. In America, the jaguar has been seen wandering from Mexico as far north as Kentucky, lat. 37° N., and in the opposite direction as far as 42° S. in South America, — a latitude which corresponds to that of the Pyrenees in the northern hemisphere. The range of the puma is still wider, for it roams from the equator to the Straits of Magellan, being often seen at Port Famine, in lat. 53° 38′ S. When the Cape of Good Hope was first colonized, the two-horned African rhinoceros was found in lat. 34° 29′ S., accompanied by the elephant, hippopotamus, and hyena. Here the migration of all these species towards the south was arrested by the ocean; but if the African continent had been prolonged still farther, and the land had been of moderate elevation, it is highly probable that they might have extended their range to a greater distance from the tropics." *

These interesting and highly important facts are brought forward in support of the erroneous impression that the extinct mammals of the drift might have inhabited the regions where their remains occur, at a time when the mean annual temperature was considerably lower than it is at present.

The living representatives of the genera to which they belong are now almost invariably inhabitants of the warm latitudes of the earth; and it is evident, therefore, that a gradual change from a colder to a warmer climate would have been favorable, rather than otherwise, to them. If, then, in those northern latitudes the conditions determining the existence of these animals, from that time to the present, have been undergoing a modification favorable to them, why are not their descendants still found there? Why have these species been completely exterminated? And if this large amount of extinction has proceeded from other than climatal causes, why are not their vacant places now filled by other analogous forms, in accordance with the established laws of nature? The facts appear to be clearly irreconcilable with the hypothesis. If, on the other hand, we view them in connection with the idea of a climate exhibiting the diversity of seasons which must follow a large degree of obliquity, we immediately experience the consciousness of having entered upon the right line of inquiry. The facts show that all, or at least nearly

* Prin. of Geol., vol. i. p. 176.

all animals are capable of adapting themselves to great extremes of heat and cold, provided adequate supplies of food can be obtained by them throughout the year. The variety and abundance of the fauna of any given locality are, therefore, directly determined by the supply of nutritious plants it furnishes, and only indirectly by climatal conditions as affecting the development of such plants. If, therefore, we conceive of tropical summers occurring in high latitudes, producing a vegetation more or less analogous to that now characteristic of southern climes, together with an essentially tropical fauna, so modified, as will hereafter be seen, in structure and habit, as to be able to endure severe cold, the seeming anomaly of a southern fauna inhabiting countries situated within the polar circles, or other localities subject to severe winters, receives a rational and complete explanation.

It does indeed, at first sight, appear impossible that these animals, typical as they are of the tropics, should be able to live through winters such as must ensue from the total absence of the sun. Allowing them able to withstand the cold, where could they have found appropriate food? The facts developed as we proceed, taken in connection with such agencies as migration, hibernation, and the capacity of taking on, in summer, stores of adipose material for winter use, will, it is thought, be sufficient to show how the herbivorous tribes were carried through. Having provided for them, we need not concern ourselves regarding the carnivorous genera.

The acuteness of observers in availing themselves of seemingly trivial chance incidents, has enabled us to verify the proposition that the outward form and osteological structure of a living animal may closely resemble that of its extinct prototype, at the same time that the habits and physiological peculiarities of the two may differ widely. In certain parts of Siberia, in which country the fossil remains of the mammoth, though widely spread over Europe and North America, are found in the greatest profusion, not only the bones, tusks, and teeth, but also the soft parts of the structure, sometimes the entire carcasses of those animals, have been found in a wonderful state of preservation, imbedded in ice and frozen mud.

"In 1772, Pallas obtained from Wiljuiskoi, in lat. 64°,

from the banks of the Wiljui, a tributary of the Lena, the carcass of a rhinoceros (*R. tichorhinus*), taken from the sand in which it must have remained congealed for ages, the soil of that region being always frozen to within a slight depth of the surface. This carcass, which was compared to a natural mummy, emitted an odor like putrid flesh, part of the skin being still covered with short crisp wool and with black and gray hairs. In allusion to the quantity of hair on the foot and head conveyed to St. Petersburg, Pallas asked whether this animal might not have inhabited a cold region of Middle Asia, its clothing being so much warmer than that of the African rhinoceros.

"Professor Brandt, of St. Petersburg, in a letter to Baron Alex. Von Humboldt, dated 1846, adds the following particulars respecting this wonderful fossil relic: — 'I have been so fortunate as to extract from cavities in the molar teeth of the Wiljui rhinoceros a small quantity of its half-chewed food, among which fragments of pine leaves, one-half of the seed of a polygonaceous plant, and very minute portions of wood with porous cells (or small fragments of coniferous wood), were still recognizable. . .'

"Thirty years after the discovery of the rhinoceros by Pallas, the entire carcass of a mammoth was obtained in 1803, by Mr. Adams, much farther to the north. It fell from a mass of ice, in which it had been encased, on the banks of the Lena, in lat. 70°; and so perfectly had the soft parts of the carcass been preserved, that the flesh, as it lay, was devoured by wolves and bears. This skeleton is still in the museum of St. Petersburg, the head retaining its integument and many of the ligaments entire. The skin of the animal was covered, first, with black bristles, thicker than horse-hair, from twelve to sixteen inches in length; secondly, with hair of a reddish-brown color, about four inches long; and thirdly, with wool of the same color as the hair, about an inch in length. Of the fur, upwards of thirty pounds' weight were gathered from the wet sandbank. The individual was nine feet high and sixteen feet long, without reckoning the large curved tusks: a size rarely surpassed by the largest living male elephants.

"It is evident, then, that the mammoth, instead of being naked, like the living Indian and African elephants, was enveloped in a thick shaggy covering of fur, probably as impenetrable to rain and cold as that of the musk-ox. The species may, as Cuvier observed, have been fitted by nature to withstand the vicissitudes of a northern climate; and it is certain that, from the moment when the carcasses, both of the rhinoceros and elephant, above described, were buried in Siberia, in latitudes 64° and 70° N., the soil must have remained frozen, and the atmosphere as cold as at this day. The discoveries made in 1843 by Mr. Middendorf, a distinguished Russian naturalist, and which he communicated to me in September 1846, afford more precise information as to the climate of the Siberian lowlands, at the period when the extinct quadrupeds were entombed. One elephant was found on the Tas, between the Obi and Yenesei, near the Arctic circle, about lat. 66° 30′ N., with some parts of the flesh in so perfect a state that the ball of the eye is now preserved in the Museum at Moscow. Another carcass, together with a young individual of the same species, was met with in the same year, 1843, in lat. 75° 15′ N., near the River Taimyr, with the flesh decayed. It was imbedded in strata of clay and sand, with erratic blocks, at about fifteen feet above the level of the sea. In the same deposit Mr. Middendorf observed the trunk of a larch tree (*Pinus larix*), the same wood as that now carried down in abundance by the Taimyr to the Arctic Sea. There were also associated marine shells of *living northern* species, and which are moreover characteristic of the drift or *glacial* deposits of Scotland and other parts of Europe." *

These instances, and especially the one last cited, indicate with satisfactory precision that those remains are to be referred to the later portion of the glacial period: — to a time when, from the prevalence of a difference in the method of distribution of the whole annual amount of solar heat over the earth's surface, there was far less contrast between the *mean annual temperature* of middle and high latitude, or rather of all latitude, than there is now, and also when the decrease of obliquity from a still larger former amount had so far tended to equalize the seasons in those northern locations, that the decreasing heat of the

* Prin. of Geol., vol. i. p. 181.

summers had become insufficient to thaw the ground to any considerable depth, or entirely liquefy the masses of snow and ice accumulated during the winters. It can be scarcely necessary to note the significance of the fact of these animals abounding in the vicinity of the pole during the glacial period. It certainly cannot consistently be construed so as to sustain the theory of an absolute relative decrease in the temperature of all latitudes of the earth — of a "Cosmic" or "Geologic Winter."

After informing us that similar remains are found imbedded in cliffs of frozen mud and ice, on the east side of Behring's Straits, in Eschscholtz Bay, lat. 66° N., in Alaska, Lyell continues: "In 1866, in the flat country near the mouths of the Yenesei, between lat. 70° and 75° N., many skeletons of mammoths were found retaining the skin and hair. The heads of most of them are said to have been turned towards the south. So late as 1869–70, an exploring expedition was made by Herr Von Maydell, under the direction of the Academy of St. Petersburg, to the river Indigiska, to examine some remains said to have been discovered there. We learn from M. Brandt that the travellers found the skin and hair as well as the bones of the *Elephas primigenius* at two points on the river, about thirty miles distant from each other, and sixty-six miles from the Arctic Sea." *

These remains of the mammoth and its associates, which are to be met with in more or less abundance from the extreme northern limits of man's explorations to as far south as Rome in Europe, and the Gulf of Mexico in North America, indicate within the period of the drift conditions highly favorable to the development of animal life. The chief prerequisite to this end is an ample supply of food for the herbivorous tribes; and with that in abundance, it may safely be inferred that the known laws of variation would superinduce such modifications in all as might be needed to adapt them to the peculiar circumstances surrounding them. We behold in these remains, therefore, the relics of animals which, like their living representatives, were essentially tropical or sub-tropical in constitution, and which, during the summers of those times,

* Prin. of Geol., vol. i. p. 183.

flourished in the enjoyment of a warmth and abundance such as their descendants now experience in the jungles of the south; yet, as we have accidentally learned, so modified in being provided with a heavy, warm covering of hair and wool combined, and also in being rendered capable of subsisting on the leaves and twigs of evergreen and other trees, during the suspended growth of more succulent and nutritious vegetation, together with habits of migration, and perhaps other unknown means, as to be able to live through the severe winters.

In relation to the capacity of these ancient animals to subsist on the kind of food above indicated, Lyell tells us that Dr. Fleming "had hinted 'that the kind of food which the existing species of elephant prefers will not enable us to determine, or even to offer a probable conjecture, concerning that of the extinct species.' No one, he said, acquainted with the gramineous character of the food of our fallow-deer, stag, or roe, would have assigned a lichen to the reindeer." * And this suggestion was made long before the discovery, by Brandt, of the fossil pine leaves and woody fibre in the molar of the Siberian rhinoceros.

On the authority of Professor Owen we are further informed, "that the teeth of the mammoth differ from those of the living elephants, whether Asiatic or African, having a larger proportion of dense enamel, which may have enabled it to subsist on the coarser ligneous tissues of trees and shrubs. In short, he is of opinion, that the structure of its teeth, as well as the nature of its epidermis and coverings, may have made it 'a meet companion for the reindeer.'"

In endeavoring to account for these remains being found in locations so far to the north, Lyell supposes they may have been transported thither from more southern latitudes by river currents. The great rivers of Siberia flowing from south to north, from temperate to arctic regions, "are all liable," he says, "like the Mackenzie, in North America, to remarkable floods, in consequence of flowing in this direction. For they are filled with running water in their upper or southern course when still frozen over for several hundred miles near their mouths, where they remain blocked up by ice for six months in every

* Prin. of Geol., vol. i. p. 185.

year. The descending waters, therefore, finding no open channel, rush over the ice, often changing their direction, and sweeping along forests and prodigious quantities of soil and gravel mixed with ice. Now the rivers of Siberia are among the largest in the world, the Yenesei having a course of 2,500, the Lena of two 2,000 miles; so that we may easily conceive that the bodies of animals which fall into their waters may be transported to vast distances towards the Arctic Sea, and, before arriving there, may be stranded upon and often frozen into thick ice. Afterwards, when the ice breaks up, they may be floated still farther towards the ocean, until at length they become buried in fluviatile and submarine deposits near the mouths of rivers." *

If these relics of glacial times occurred only in the above described situation, — in fluviatile deposits of large rivers flowing from the south to the north, and there only in very rare, exceptional instances, — this hypothesis might, perhaps, be considered more or less probable. Such, however, is far from being the case; for fossils of this class are found in greater or less abundance, in all conceivable situations, throughout the northerly portions of both the eastern and western continents of the northern hemisphere; and this fact alone releases us from the necessity of having recourse to Lyell's improbable supposition, and leads to the natural and consistent view that the animals in question were denizens of all the regions where their remains are found, and that those regions, even if they may hereafter be found to extend to the pole itself, were in palæolithic and glacial times, in climatic conditions, vegetable productions, and so forth, specially adapted to their needs.

Lyell asserts on page 184, what is undoubtedly true, "that the ice or congealed mud in which the bodies of such quadrupeds were enveloped, has never once been melted since the day when they perished, so as to allow the free percolation of water through the matrix; for had this been the case, the soft parts of the animals could not have remained undecomposed."

Without venturing any suggestion as to the origin of the marine shells present in the river Taimyr deposits, it is evidently impossible that they, or any other of the glaci-

* Prin. of Geol. vol. i. p. 186.

ated formations containing undecomposed portions of soft animal tissue, could have remained for any considerable time under the waters of the sea without attaining a temperature sufficiently high to insure their liquefaction, and the consequent decomposition of the imbedded carcasses; and it may be considered reasonably certain that the strata in question, since their deposition, have never been submerged beneath the water of the ocean.

Corroborative of this view, and at the same time affording light in regard to the peculiarities of the prevailing climate of those times, are the peculiar states or conditions in which the Siberian remains are found. The dried-up, mummy-like appearance of the Wiljui rhinoceros may be supposed to afford a correct indication of the state of the atmosphere at the time of the animal's demise. It is equally impossible either that the carcass could have been immediately frozen up and enveloped in its icy covering, in which case the flesh would have been preserved in the fresh state characterizing other specimens which were undoubtedly so frozen and enveloped, or that it could have been subjected to the influence of a warm and moist atmosphere, such as would be favorable to the decomposition of animal substances. The peculiar state of preservation in which it was found indicates with precision a dry and antiseptic quality of air like that now observable in certain countries; a condition of the atmosphere likely to ensue during the latter part of the summers of those times, from the long continuance of the sun above the horizon. So, also, the partially decomposed state of other specimens — a condition that must have preceded congelation and inhumation — points with the same certainty to the fact that a moist and warm state of atmosphere prevailed at the time and place of the death of those individuals.

The rigorous cold incident to the winters of drift and glacial times, is directly and convincingly shown by the indelible traces left upon the rocks by ice; and although no evidence of a strictly analogous character can be adduced to indicate the prevalence of alternating hot summers, that which we can command, although less direct, is every whit as positive and satisfactory. The former presence, in great number, of herbivorous animals of large size, belonging to species which require a high temperature for their development, in regions where now the prevailing climate renders impossible a vegetation capable of supplying them with food, and which during no portion of the year exhibits the necessary conditions for the development of those classes of forms, is sufficient to authorize the conclusion that when they inhabited such regions, some part of the year at least must have exhibited the conditions of life on which their existence may naturally be supposed to depend. The first and most important of these conditions is a high temperature, the others being more or less directly dependent upon it. We are absolutely certain that the winters were cold, and equally so that the animals in question were provided with an adequate protection against such cold; although, unlike the reindeer and polar bear, incapable of enduring a continuously low temperature. For, were that not the case, their descendants would now be found occupying the same territory. It is thus evident that they could not exist except through the intervention of annual seasons of warmth; and their presence, therefore, is positive proof of such seasons, and is in direct contravention to the idea of continuous cold, such as must result from an absolute decrease in the terrestrial temperature.

Geologists recognize in the drift traces of the hot summers, and Lyell attributes them to intercalated periods of warmth of comparatively long duration. As confirmatory of the opinion that the men of the early stone age had often to contend with a climate more severe than that now prevailing in the same parts of Europe, after having stated that at Fisherton, near Salisbury, in England,* "one of the rude flint implements of the earliest stone age was found in drift containing the mammoth and Siberian rhinoceros, together with the Greenland lemming and a *Spermophilus*, another northern form of rodent allied to the marmot, besides the tiger, hyena, horse, and other extinct and living species," he continues: "But we find in some parts of the drift evidence of a conflicting character, such as may suggest the idea of the occasional intercalation of more genial seasons of sufficient duration to allow of the migration and temporary settlement of species coming from another and more

* Prin. of Geol., vol. i. p. 191.

southern province of mammalia, so that their remains were buried in river gravels at the same level as the bones of animals and shells of a more northern climate. . . . Bones of the hippopotamus, of a species closely allied to that now inhabiting the Nile, are often accompanied in the valley of the Thames and elsewhere, by a species of bivalve shell, *Cyrena* (*Corbicula*) *fluminalis*, now living in the Nile and ranging through a great part of Asia as far as Tibet, but quite extinct in the rivers of Europe. Imbedded in the same alluvium with this shell, we find at Grays in Essex, *Unio littoralis*, a mussel no longer British, but abounding in France in rivers more southern than the Thames. The *Hydrobia marginata* is also a shell sometimes met with in the drift, a species now inhabiting more southern latitudes in Europe. The kind of elephant and rhinoceros accompanying the Cyrena at Grays (*E. antiquus* and *R. megarhinus*) are not the same as the mammoth and rhinoceros which occur with their flesh in the ice and frozen mud of Siberia, or in those assemblages of mammalia which have an arctic character in the drift of England, France, and Germany. Some zoologists conjecture that the fossil species of hippopotamus was fitted for a cold climate, but it seems more probable," says Lyell, "that when the temperature of the river water was congenial to the Cyrena above mentioned, it was also suited to the hippopotamus."

The extraordinary assemblage upon intermediate ground, not only of the organisms properly indigenous to it, but those also of both arctic and subtropical latitudes, indicated in the foregoing extracts, is, we believe, wholly inconsistent with the theory of intercalated periods of warmth occurring in the midst of the Glacial epoch. On any view of the facts, it must be exceedingly difficult to believe that anything analogous to the present climate of Egypt prevailed in England for a length of time sufficient for the migration, and even the temporary establishment, of Egyptian species so far north, at a time when all the rest of the earth's surface was subject to what has formerly appeared to be a considerable reduction of the ordinary normal temperature.

The contemporaneity of these fossils of the English drift seems to be implied in their peculiar position. Lyell does not intimate but what they are scattered indiscriminately throughout the entire formation, and the natural inference is, that they are so disposed. In fact, he asserts that they occur at the same level. Now, if these, under present climatal conditions, geographically distinct classes of animals inhabited the British Isles in succession, the Siberian mammoth and rhinoceros, Greenland lemming, spermophilus, and other arctic species, flourishing for a time, being succeeded by a fauna analogous to that of the present, and this last, after a time, giving place to another, such as is now observed in latitudes twenty-five or thirty degrees farther to the south, the remains of each class would be found disposed in separate and distinct layers, and not at random throughout the entire mass; and therefore, until it shall be shown that they do occur in such separate order, we have the right to assume that the whole group of fossils are the remains of the indigenous fauna of the British Isles during the period of the drift.

It by no means follows because a more or less distinct structural difference is observable between the elephant and rhinoceros at Grays, and those specimens found at Fisherton, that they were not all specifically adapted to the same climatal conditions; for, after what we have learned of their Asiatic contemporaries, it is very easy to conceive that osteological differences might, in the cases in point, co-exist with identity of habit, epidermatic similarity, and a like exterior provision against severe cold.

The fauna of England in the drift period, then, included animals belonging to three of the present geographical classes, or types. Arctic, and subtropical species, and those intermediate between the two, occupied at one and the same time a common country as their chosen habitat. Such an assemblage is alike inconsistent either with an absolute increase or absolute decrease in mean temperature; for, if the climate, on the whole, was colder than now, the subtropical tribes could not maintain existence; and if warmer, the arctic types must have perished. Such a fauna, on the other hand, may be supposed perfectly adapted to a climate exhibiting summers of tropical aspect, in which the southern species would enjoy the peculiar conditions of life suited to their wants; such summers alternating with winters that should fulfil the requirements of the arctic types; the whole fauna being so far modified in structure and habit from present forms, as to be able to maintain existence during those portions of the year to which they may now, to us, seem more or less specifically unadapted. From this stand-point, the facts tending to show the prevalence of warmth during the drift and glacial eras, instead of leading us through the forced exercise of the imagination into vagaries more or less absurd, find an appropriate place in the superstructure of a just system, and we can understand how a *modified* hippopotamus and Cyrena were able to co-exist with the Siberian mammoth and rhinoceros, and Greenland lemming, among mammals, and the Limnea among mollusks.

If we accept as probable the conclusion that the mammoth and its associates were denizens of, and able to procure subsistence for themselves throughout the year within regions now, at all seasons, presenting the appearance of desolate and deserted icy wastes, by means of a climate then prevailing, consequent on a large degree of inclination of the earth's axis, entailing great annual extremes of heat and cold; questions present themselves relative to the effect of such extremes on vegetation, as affecting the supplies of food for those animals.

Without attempting, at this time, any extended discussion of points involved, it may be sufficient to observe that the power of the perennial vegetation of temperate climes, the seeds and spores of annual varieties of the same, and, in very many instances also, the seeds and spores of subtropical and tropical species to withstand the effects of cold, has, as far as now known, no limit. It is very probable that explorers in the vicinity of the poles have experienced, approximately, the extreme degree of terrestrial cold, which may be stated at about –70° or –80° F. If this be a tolerably correct estimate of the excesses of cold of polar winters, alternating with seasons altogether unworthy of being designated as summer, it is plain that when such polar winters should be succeeded by summers equalling them in duration, which summers should, for a portion of the time, enjoy the effects of the vertical rays of the sun, with no intervention of night, a considerable reduction from the degree of rigor at present observed may safely be allowed. The effect of an excess of cold on the flora of any given locality can be but slightly, if indeed, in any degree, affected by the duration of such excess; for, after such flora shall have assumed its winter status, a fatal degree of cold would perform its work upon it as effectually in a few hours as in as many days or months, and whatever amount of rigor it could withstand unharmed for a few hours it could also endure for a whole winter. The frequent recurrence of brief periods of excessive cold during the winters of the northwestern states of America and other exposed locations,—the thermometer often indicating –40° and –50° F., and even still greater degrees of cold, with no observable detrimental effect upon the natural vegetation of such places, is quite sufficient to demonstrate the ability of the floras of temperate latitudes to withstand winters approaching in severity those of polar regions.

Were we to admit, however, what there is not the slightest necessity for doing, that excessively cold winters would have a tendency, in the case in point, to produce a stinted and scant flora, we still have the authority of Darwin, as cited by Lyell, that even a vegetation of this character is sufficient to sustain a large amount of animal life.* "It has often been taken for granted that herbivorous animals, of large size require a very luxuriant vegetation for their support; but this opinion is, according to Mr. Darwin, completely erroneous:—'It has been derived,' he says, 'from our acquaintance with India and the Indian islands, where the mind has been accustomed to associate troops of elephants with noble forests and impenetrable jungles. But the southern parts of Africa, from the tropic of Capricorn to the Cape of Good Hope, although sterile and desert, are remarkable for the number and great bulk of their indigenous quadrupeds. We there meet with an elephant, five species of rhinoceros, a hippopotamus, a giraffe, the *Bos Caffer*, the elan, two zebras, the quagga, two gnus, and several antelopes. Nor must we suppose that, while the species are numerous, the individuals of each kind are few. Dr. Andrew Smith saw, in one day's march, in lat. 24° S., without wandering to any great distance on either side, about 150 rhinoce-

* Prin. of Geol., vol. i. p. 189.

roses, with several herds of giraffes, and his party had killed, on the previous night, eight hippopotamuses. Yet the country which they inhabited was thinly covered with grass and bushes about four feet high, and still more thinly with mimosa-trees, so that the wagons of the travellers were not prevented from proceeding in a nearly direct line.'

"In order to explain how so many animals can find support in this region, it is suggested that the underwood, of which their food chiefly consists, may contain much nutriment in a small bulk, and also that the vegetation has a rapid growth; for no sooner is a part consumed, than its place, says Dr. Smith, is supplied by a fresh stock."

From all of the foregoing facts Lyell infers that in the time of the mammoth and of European drift and cave deposits, which time, if not actually an integral portion of the Glacial period, must have been situated upon its immediate confines, constituting, in fact, an era of transition from the climatal excesses of the ice period to the comparative uniformity of later times, "a large region in Central Asia, including, perhaps, the southern half of Siberia, enjoyed, at no very remote period in the earth's history, a climate sufficiently mild to afford food for numerous herds of elephants and rhinoceroses, *of species distinct from those now living.*" * He admits that elephants and rhinoceroses would find it impossible to subsist, at the present time, even in South Siberia; and therefore, according to his theory, it must have been considerably warmer, in the time of the mammoth, wherever that animal ranged, than at present. But as we proceed, we find that contemporaneous European deposits indicate a temperature, "according to Mr. Prestwich, 20° Fahrenheit colder than now, or such as would now belong to a country from 10° to 15° of latitude more to the north." †

At a time, therefore, when the climate of "Central Asia, including, perhaps, the southern half of Siberia" — an area included between the forty-fifth and sixtieth parallels of north latitude — was considerably warmer than now, the climate of that portion of Europe comprised within the same limits was twenty degrees colder than now; and

* Prin. of Geol., vol. i. p. 188. † Ibid. 190.

our assent is required to the belief that the causes assigned by Lyell to account for former climatic variations, hereafter to be considered, operated, during the era in question, to produce, at different places situated in the same latitude, a difference in mean annual temperature equal at least to 40° F. This amount of difference is sufficiently extraordinary to test our credulity to its utmost extent when predicated of points that might happen to be subjected to the extremes of the various local influences affecting climate; but when it is sought to make it general, — to extend it over the major portions of two vast continents, — there seems to be no rational alternative but to reject it as entirely inadmissible. Opposed also to Lyell's theory, which endeavors to assign a limit to the northerly range of the mammoth and contemporaneous animals, is the fact that their remains are to be found in the eastern hemisphere, as far north as the seventy-fifth parallel, and on the American continent, up to the sixty-sixth degree of latitude, and, if the land extended so far in those times, inferentially to the pole itself; showing, without doubt, that the peculiar climatic conditions upon which depended the existence of those species were the product, not of inferior local agencies, but rather of some great primary cause, whose effects were relatively the same in all latitudes of the northern hemisphere, and, we may naturally conclude, over the whole world. This consideration, in connection with the identical nature of the palæolithic and glacial deposits, is wholly at variance with our author's hypothesis, and is only consistent with that which regards them as effects of a grand primal agency, operating uniformly, and producing the same relative effects, not over a small portion only, but over the whole surface of the earth.

How perplexing the effort to bring the facts we have cited into harmony either with the hypothesis just considered, or the unsubstantial and visionary assumption of a period of absolute excessive cold, pervading alike the whole earth, — a cosmic winter, enduring for indefinite ages, caused by some unknown, mysterious, extrinsic agency; transforming the teeming and fruitful world, from equator to the poles, to a bleak and dreary ball of ice, revolving in darkness and desolation, and devoid of life, — and necessitating, at its close, a re-creation of all the varied forms of life by which it had previously been inhabited! For such, indeed, is the picture that has been presented to us of the earth during this period. How much more consonant with the general plan — a plan founded not on fitful and capricious exertions of creative energy, but upon immutable natural laws, in the system, which, in the reduction of observed natural phenomena, avoids the necessity of violent and sudden changes, catastrophes, and cataclasms in the inorganic world, involving complete extinctions and subsequent re-creations in the organic, and seeks rather to evolve a principle under which the general course of nature shall continue uninterruptedly the same throughout all the ages of the world.

CHAPTER II.

EVIDENCE SHOWING THE PREVALENCE OF DIVERSE CLIMATAL CONDITIONS IN FORMER TIMES.—LYELL'S RETROSPECT, CONTINUED.

THE PLIOCENE GEOLOGICAL PERIOD NOT YET COMPLETED.—COLD OF PLIOCENE GLACIAL INTERVAL EXAGGERATED.—METEOROLOGICAL EFFECT OF EXTREME OBLIQUITY THE CAUSE OF GEOLOGICAL REVOLUTIONS.—PROFESSOR TYNDALL ON THE GLACIAL PERIOD.—SUPERFICIAL CHANGES OF THE GLACIAL PERIOD.—INTER-GLACIAL PERIOD AND EARLIER PLIOCENE.

UP to the present time the prevailing opinion seems to have been that the geological revolution involved in the Pliocene epoch was completed previous to the commencement of the latest Glacial period, and that such epoch constituted the last of the third or Cainozoic great series of similar revolutions, into which the geological history of the earth has been divided. Subsequent time—Post Pliocene and Recent, so called, with their subdivisions—seems to have been regarded, in a somewhat indefinite way, as the inauguration, not only of a new geological period, but also of another grand series of like periods.

As the geologist, in his peculiar line of investigation, recedes further and further into the past, with nothing but random conjecture to guide him in his estimates, it is natural to suppose that he must be less and less able to correctly appreciate the amounts of time embraced within its various divisions. He will be found extremely liable, perhaps, to allow, as the duration of what may, in reality, constitute but a mere fraction of a recent epoch, a lapse of time as great as that which he assigns to a complete epoch further removed from his time. Confusion and inaccuracy must attend his efforts to reduce to chronological order the imperfect rocky records it is his province to decipher. Under the new system, however, we exchange conjecture for an infallible guide, by means of which a terrestrial chronology becomes possible, as exact as the mutilated and broken annals upon which it must necessarily rest will admit.

One of the most manifest errors to which our attention is directed under our new theory, is that which assumes the Pliocene age, as classified by geologists, to embrace a complete geological revolution. And to guard against ambiguity in the mind of the reader, it may in this connection be advisable to say that geological revolutions or periods, cycles of primary climatal change, and half transverse revolutions of the earth, are convertible terms — a proposition embodying, perhaps, the most concise statement of the new theory.

As we continue our retrospect, we shall see that unmistakable traces exist of an *apparent* gradual decrease in the terrestrial temperature, from the close of Miocene times, when a uniformity of seasons and an appearance of abnormal warmth prevailed, resulting from the position of the earth with regard to the sun, at the time of equatorial coincidence with the ecliptic, down to the time of the greatest *apparent* cold of the Glacial period; this last climatic state being the result of an extreme inequality of seasons entailed by the position the earth had then assumed, the equator having become perpendicular to the orbital plane. This portion of time, corresponding to one half of the semi-transverse revolution, includes all of what has been considered as constituting the Pliocene period. We have already seen that the geological record indicates an *apparent* increase of warmth through all the ages intervening between glacial and modern times; an appearance consequent on the gradual return of the seasons to the former state of uniformity. But even this last named portion of the semi-transverse revolution, embracing Post Pliocene and Recent times, added to the before stated quantity, does not make up the whole sum of 180° of circular motion, the amount required to include a complete cycle of climatal change; for if we take as the commencement of the geological period, the point of equatorial coincidence with the ecliptic, it will require the lapse of one hundred and seventy-six thousand years from the present time, as we reckon, to complete the cycle of changes, and to bring the earth into the same position with regard to the sun it occupied at the commencement. If, however, we take, as a more proper beginning, the point at which increasing obliquity had attained a value of 45°, and where the meteorological effects of the increased inequality of seasons had begun to effect, in a marked degree, the various changes incident to the glacial intervals of such periods, we find that over a half million years will yet be required to complete the Pliocene geological epoch, and inaugurate the Ice period, or interval of change, of its successor. This immense amount of time which must, to a great extent, remain a blank page in the record of the future geologist, will constitute an era of repose, during which the superficial aspect of the earth will remain comparatively unchanged, except in so far as it may be affected by those upheavals and subsidences depending on subterranean or other forces independent, as far as now known, of the chain of causation we are endeavoring to follow. At the close of this era of repose, the rough harrow of the succeeding Ice period will rasp down the present terrestrial surface, covering it with piles of new material. Then will the drift deposits of the last ice interval (the phenomena of which will next receive our attention) become conformable and recognizable as the uppermost member of the completed Pliocene great series.

The Glacial Interval of the Pliocene Period.—The most authentic and startling revelation of former vicissitudes in the climate of the world, is to be found in the monuments constituting the geological record of the Pliocene Glacial interval. We are not, however, to assume from this fact that the cold of this era exceeded in rigor that which characterized its predecessors; its peculiar phenomena being more legible and more abundant than that of the others only from being more recent, and because it has not, like them, been subjected to the destructive influences of subsequent periods of change, and for the further reason, also, that a large part of it remains exposed to view upon the present surface of the earth.

Among the monuments or glacial phenomena of the above-mentioned interval are, first, extensive deposits of loose material, sometimes bearing more or less distinct traces of stratification, known as *drift;* secondly, piles of rocky debris, denominated moraines, deposited by and determining the terminal and other boundaries of ancient glaciers which in extent largely exceeded those of the present time; and, thirdly, erratic boulders, or large insulated masses of rock, occurring far from the original parent ledges of the same, some of these, according to Lyell, being "polished and striated on one or more of their sides, in a manner strictly analogous to stones imbedded in the moraines of existing glaciers in the Alps," and the underlying solid rocks, in many instances, being "marked by similar scratches and rectilinear furrows, their direction usually coinciding with the course which the erratics themselves had taken." *

It was at first supposed that these appearances extended from the north only to intermediate latitudes of the temperate zones; but further exploration, both in the northern and southern hemispheres, succeeded in tracing them further and further towards the tropics, until, at last, Professor Louis Agassiz, in the endeavor to substantiate the hypothesis that the Glacial period was an era characterized by a large absolute decrease in terrestrial temperature,—a "geologic" or "cosmic winter,"—sought for, and discovered, in tropical South America, evidence that this class of phenomena extended almost, if not quite to the equator.

There is little doubt that considerable exaggeration has attended the discovery and interpretation of ancient glacial phenomena; but, after making due allowance for the same, it must be conceded that the nature of the facts are such as to give an air of probability to the theory that an absolute decrease from the normal temperature of the world prevailed at that time.

The opinions to which naturalists have arrived in regard to the intensity of the cold of the glacial period may be inferred from the quotations following. Says Darwin, "We have evidence of almost every conceivable kind, organic and inorganic, that within a very recent geological period, Central Europe and North America suffered under

* Prin. of Geol., vol. i. p. 192.

an Arctic climate. The ruins of a house burnt by fire do not tell their tale more plainly, than do the mountains of Scotland and Wales, with their scored flanks, polished surfaces, and perched boulders, of the icy streams with which their valleys were lately filled. So greatly has the climate of Europe changed, that in Northern Italy, gigantic moraines, left by old glaciers, are now clothed by the vine and maize. Throughout a large part of the United States, erratic boulders, and rocks scored by drifted icebergs and coast ice, plainly reveal a former cold period." *

Professor Agassiz, an indefatigable student of glacial phenomena, ancient and modern, furnishes us, in the citation following, with the means of arriving at his idea of the degree of cold which characterized the Glacial period: "To this cosmic winter, which, judging from all the phenomena connected with it, may have lasted for thousands of centuries, we must look for the key to the geological history of the Amazonian valley. I am aware that this suggestion will appear extravagant. But is it, after all, so improbable that when Central Europe was covered with ice thousands of feet thick; when the glaciers of Great Britain ploughed into the sea, and when those of the Swiss mountains had ten times their present altitude; when every lake in Northern Italy was filled with ice, and these frozen masses extended even into Northern Africa; when a sheet of ice reaching nearly to the summit of Mount Washington (that is, having a thickness of nearly six thousand feet) moved over the continent of North America, — is it so improbable that, in this epoch of universal cold, the valley of the Amazons also had its glacier poured down into it from the accumulation of snow in the Cordilleras, and swollen laterally by the tributary glaciers descending from the tablelands of Guiana and Brazil?" † His conclusion is, therefore, that during the Glacial period, which he correctly conjectures to have covered hundreds of thousands of years, the terrestrial temperature became reduced to such an extent that, under the direct influence of the sun's vertical rays as exhibited at the equator, the valley of the river Amazon became filled with an immense glacier — a mass of ice several thousand miles long, from five to

* Origin of Species, p. 319.

† Journey in Brazil, p. 425.

seven hundred miles in width, and thousands of feet in thickness. With such stupendous processes of glaciation in progress at the equator, it is easy to conceive what must have been the climatal condition of other latitudes.

An annual average temperature of 32° F. is required for the formation of glaciers; and, according to Agassiz, "a degree of temperature in the annual average of any given locality corresponds to a degree of latitude; that is, a degree of temperature is lost for every degree of latitude as we travel northward" (or from the equator), "or gained for every degree of latitude as we travel southward" (or towards the equator). If, therefore, during the Glacial epoch, the position of the earth with regard to the sun was the same as now, involving the same differences in temperature relative to latitude, when the mean average at the equator was 32°, the annual mean at Alexandria, in Egypt, and Savannah, in Georgia, would be 0°; at Rome, –10°; at London –20°, and so on towards the pole. These figures denote not the cold of the winters, but the average temperature of the whole year. Under these circumstances, the temperature of by far the larger part of the world would approach that of celestial spaces far removed from the influence of bodies analogous to our sun. The fluid portion of the globe would, in a comparatively short time, become solid, and all organic life be at once annihilated.

Now, if we attempt to account for the extreme degree of refrigeration above indicated, there seems to be but two hypotheses available. Either the earth must have abandoned its usual orbit, and, for the time being, described another situated somewhere upon the outermost confines of the solar system, or the sun must have ceased almost entirely to impart the ordinary supplies of light and heat, not only to the earth, but to all the other planets dependent upon it; for if the sun continued to emit the normal quantity, the only method by which the annual supply at the earth could suffer great diminution would be to largely increase the distance between the two; and, conversely, if the mean distance be supposed to have continued always the same, any considerable reduction in the whole annual amount of light and heat received by the earth must, of necessity, be consequent upon the neutralization of the solar influence. Both of these propositions we hold to be equally absurd. It seems wholly unnecessary to assert that the earth has never, from the beginning, suffered under the effects of so excessive a degree of cold; and, although the advocates of the cosmic winter theory have assented to the notion of a universal extinction of life during the Ice period, as an unavoidable corollary of their doctrine, the geological record shows that no such extinction has ever taken place.

"We have found evidence," says Lyell, "that most of the testacea, and not a few of the quadrupeds, which preceded, were of the same species as those which followed the extreme cold. To whatever local disturbances this cold may have given rise in the distribution of species, it seems to have done little in effecting their annihilation." *

Agassiz found on the banks of the Solimoens, in the Amazonian valley, in a position beneath his imaginary glacier, the remains of "a vegetation similar in general character to that which prevails there to-day." † And although, for obvious reasons, organic remains are scarce, and sometimes entirely wanting in drift deposits, enough are found therein to warrant the assurance that all the various forms of life maintained existence during the almost interminable ages of the Glacial period. Throughout this immensely long era, as estimated in years, the general course of life over the whole earth seems to have been continuous and uninterrupted, exhibiting only the gradual and uniform change everywhere observable throughout the whole geological record, with no greater amount, perhaps, of modification and extinction than may be traced in other like periods of the world's history. This continuity of life is evidently incompatible with a cosmic winter as long and as rigorous as the one we have contemplated; and we may rest with perfect confidence in the conclusion that no such winter has ever occurred in the terrestrial annals since the earth first assumed its present relations in the solar system.

It does not follow, however, because we are compelled to dissent from Professor Agassiz' estimate of the degree of cold indicated by the phenomena he witnessed in Brazil,

* Prin. of Geol., vol. i. p. 306.

† Journey in Brazil, p. 424.

that we are to reject his statements as competent authority in relation to the character of those facts.

We cannot for an instant suppose that this distinguished naturalist and keen observer, who for many years has made the study of both ancient and modern glacial phenomena a specialty, can have been mistaken as to the nature of the facts which came under his observation in Brazil. There is, perhaps, no man living whose experience renders him more capable of identifying the traces of glacial action than he. Admitting his ability in the premises, — allowing him able to recognize a moraine, to tell an erratic boulder from rock in place, or to distinguish between drift material and that of other formations, — there is no method by which we can avoid the conclusion, on a review of the results of his labors in tropical South America (the importance of which subject in this connection leads us to reserve its consideration for a separate chapter), that in Glacial times processes were going on, at very moderate elevations, which at present are unknown there; being now confined to high latitudes, and to the more elevated regions of mountainous countries. It may be sufficient to observe in this place, that the facts as narrated, although far from indicating the prevalence of an extreme degree of cold, establish with sufficient accuracy, that formerly the climate there differed altogether from that of the present time, approaching much more nearly, in many respects, that of temperate latitudes. But even this modification of climate at the equator is entirely inconsistent with the present status; for it is wholly impossible that a system of glaciation could have been in progress there at moderate heights, under present conditions; that is, with the sun's rays vertical, or nearly so, throughout the entire year. This limited amount of glaciation, this prevalence of a temperate climate at the equator, is as rationally and satisfactorily explained under our system, and is as conformable to the relative positions we assign to the earth and sun at that time, as are all the other phenomena of the Glacial period, as exhibited in other parts of the earth. The position of the earth —its axis of diurnal rotation coinciding with the orbital plane — would have the effect to produce, in equatorial regions, two short and cold winters, and two short and hot summers in each year; and it is easy to imagine that under the meteorological complications incident to such an arrangement of the seasons, little time would suffice to inaugurate systems of glaciation in elevated regions, such as should pile up moraines, distribute erratics, and, perhaps, fill a possible Amazonian lake or sea with melting bergs and fields of ice laden with the disintegrated material of the mountains in which they originated.

The Glacial period, then, when viewed from the proper stand-point, so far from bearing the appearance of the cosmic or geologic winter before described, exhibits rather the phenomena that must result from a more equal distribution of the whole annual amount of solar heat over the earth's surface than now obtains; the traces of cold, of unusual glaciation, being not the consequence of a diminution of the solar energy, but the effect rather of the great contrasts of seasons entailed by such more equal distribution, as effected by the peculiar position of the earth with regard to the sun prevailing at that time.

Meteorological Effect of Extreme Obliquity the Cause of Geological Revolutions. — From a general survey of phenomena, geologists have made a division of past time into periods of revolution and tranquillity — of convulsion and repose. The several series of changes or revolutions in the earth's crust seem to have been effected within stated seasons or periods, and these periods appear to alternate with others, in the records of which no trace of such changes is to be found. The appearances on which the doctrine of alternate periods of disorder and repose is founded, are alluded to by Lyell as follows: * "It has been truly observed, that when we arrange the fossiliferous formations in chronological order, they constitute a broken and defective series of monuments: we pass without any intermediate gradations from systems of strata which are horizontal, to other systems which are highly inclined — from rocks of peculiar mineral composition to others which have a character wholly distinct — from one assemblage of organic remains to another, in which frequently nearly all the species, and a large part of the genera, are different. These violations of continuity are so common as to constitute in most

* Prin. of Geol., vol. i. p. 298.

regions the rule rather than the exception, and they have been considered by many geologists as conclusive in favor of sudden revolutions in the inanimate and animate world. We have already seen that, according to the speculations of some writers, there have been in the past history of the planet alternate periods of tranquillity and convulsion, the former enduring for ages, and resembling the state of things now experienced by man; the other brief, transient, and paroxysmal, giving rise to new mountains, seas, and valleys, annihilating one set of organic beings, and ushering in the creation of another." He then proceeds to show that these theoretical views in relation to the manner in which the changes in question were brought about, "are not borne out by a fair interpretation of geological monuments." But when, in order to dispense with sudden and catastrophal "revolutions in the geological order of events," he endeavors, as he does, to assign a potency to the acknowledged agents of change, when in a state of repose, as now seen, such as they can possess only in a condition of greatly increased activity, the difficulties under which he labors become at once apparent. If, at the present time, which, in the foregoing quotation, he recognizes as a period of tranquillity, these agents are exhibiting a degree of activity sufficient to effect the mighty changes in question, if the enormous denudations and depositions to which the geological record bears witness are now progressing at as rapid a rate as at other periods of the world's history, this fact alone is quite sufficient to overthrow and invalidate the doctrine of alternative periods of convulsion and repose — a doctrine to which all assent, and the truth of which is beyond question. It is no wonder, therefore, that the author finds it "necessary in the present state of science to supply some part of the assumed course of nature hypothetically."

There appear to be two principal forces or agents, whose office it is to effect changes in the aspect of the superficial portion of the solid substance of the earth. The first of these, known for the most part only through its effects, is that producing elevations and depressions of the earth's surface, such as are known to be now in progress in Sweden, Greenland, and other countries; and the other, that which depends on meteorological processes more or less connected with variations in the terrestrial temperature. To the former can be attributed little more than those mere oscillations of level which serve, at intervals, to raise limited portions of the surface of the land to greater or less elevations above, or depress the same to a greater or less depth below, the level of the sea; but which can only in comparatively rare instances produce, to any considerable extent, the great inequality of surface that so generally prevails over the whole world.

Under the idea that large masses of land emerge from the depths of the ocean at a bound, or as suddenly sink beneath the waves, we might be justified in supposing that the advancing or receding waters would effect considerable change in the position of the loose material upon the surface of such masses. But as it is manifest from the most exact observation and measurements that these oscillations of level proceed at a very slow rate, a few inches only of elevation or depression occurring in a century, it is evident that this agency can have effected little or nothing in producing those inequalities upon the earth's surface due to aqueous causes.

And again, when we seek, *under the present status*, to explain former superficial changes, the vast denudations and depositions, and the formation of river valleys, which have been effected through the agency of running water, by reference to the second class of causes, we cannot fail soon to realize that the attempt involves an impossibility. No person, for instance, accustomed to the critical observation of natural phenomena can examine the valley of a river, much less follow its windings from source to mouth, without receiving upon the mind a decided impression that the valley must, in some way, have been the natural consequence or work of the river. Standing on the bank of the stream, however, and turning his eyes towards the distant heights, whose summits, many hundred feet, perhaps, above the river level, mark the extreme bounds of fluviatile action, and constitute the remains and continuation of a former unbroken surface, he can no less resist the conviction that the sluggish stream at his feet, even in its seasons of greatest activity, could never, by possibility, have scooped out the immense basin. No argument is needed to assure him that the comparatively insignificant river, as he beholds it, might roll its annual flood and ebb from year to year, century to century, and from age to age, perhaps, without altering in the least the general appearance and shape of the valley.

Now, there can be no question that the river excavated the valley. Every geologist will admit this much. Neither will it be denied that while the work was in progress, the volume of the annual accumulations of water must have been largely in excess of what it is now. The stream must have been wider and deeper, and its current much more powerful; and even under these circumstances it must have required an immense interval of time to accomplish the work, the result of which in effecting superficial change is here only half apparent; for the material removed in the denudation of the valley has been deposited elsewhere, producing over some other area an equal amount of change. The swollen condition of the streams, the vast inundations of the Glacial period, and consequent superficial changes are not confined to particular localities, but are common, under various phases, to all parts of the world. Even at a distance from present watercourses, the traces of the action of water, either in the fluid state or in the form of ice, during this period, are everywhere abundant, and it is utterly impossible to account for this state of things without recourse to the hypothesis of a totally different climate from that which now prevails; and the question recurs, — how must the climate of the Glacial period have differed from that of the present time to produce the phenomena above mentioned?

The geological record, as we approach the Glacial period, *seems* to indicate a lower and lower temperature; and as the appearances of cold increase, the agents of superficial change exhibit greater and greater activity. The water of the rivers increases largely in volume, and their annual inundations assume larger and larger proportions, until at last, during the time of greatest cold, fluviatile and diluvial forces seem to have attained their extreme point of efficiency. This, it is scarcely necessary to add, involves a corresponding increase in activity of the ordinary meteorological agencies depending more or less on contrasts of temperature, such as the evaporization and condensation of moisture, and the congelation and liquefaction of the same. Now, is this acceleration of meteorological activity consistent with a correspondingly gradual absolute decrease in the terrestrial temperature?

Says Lyell, "As the temperature of the atmosphere diminishes gradually from the equator towards the pole, the evaporation of water and the quantity of rain diminish also." * Therefore, if we wish to arrive at a correct estimate of the actual results of an absolute decrease in temperature, we have but to make a journey northward. As before stated, each degree of latitude we reach will represent a diminution of one degree of temperature; and if we proceed far enough in the direction of the pole, we shall at length arrive where the cold may be considered as approximating that which any reasonable supposition can assign to the Glacial period. But in the course of our journey, although we find the cold to increase in the above definite proportion, no corresponding increase will be observable in the efficiency of those aqeuous forces which we must consider as mainly instrumental in effecting the changes incident to the glacial period. On the contrary, we shall find them acting with diminished vigor as the temperature falls; and when we arrive in the vicinity of the pole — the region of perpetual frost — nothing can be farther from the transition state, or more stable than the condition of the earth's surface as there presented, notwithstanding evidence of the former activity of these forces is quite as abundant there as in any other part of the world. If intense cold has effected through meteorological, or, indeed, any other agency, the superficial changes incident to the Glacial period, why do not explorers in polar latitudes now find analogous phenomena in progress there? Under this hypothesis, they ought to behold periodical liquefactions of the masses of snow and ice, giving rise to tremendous floods; these last denuding extensive surfaces, scooping out valleys, and transferring the material of the earth's surface from one locality to another. Such, however, is not the case. The masses of snow and ice retain substantially the solid form throughout the entire year; and of the limited amount of atmospheric moisture precipitated

* Prin. of Geol., vol. i. p. 323.

upon the land, a large proportion reaches the sea by means of glaciers, in the form of ice, and becomes liquefied only after transportation by ocean currents to more genial climes. It may therefore be asserted with perfect confidence, that the superficial changes incident to the Glacial period could not have been brought about by a mere reduction of the earth's temperature, being clearly inconsistent with, and impossible under, the conditions necessarily attendant upon such reduction.

Let us, on the other hand, imagine the effect likely to be produced by the heat of a tropical summer acting upon those polar accumulations of snow and ice. The term *tropical*, however, does not convey a just idea. To obtain that, we must conceive of a tropical noontide, continuous throughout a considerable part of the season — one blazing noon, unbroken even by the cooling influences of morning and evening, to say nothing of night. The vast denudations involving extentive transfers of material, and the cause which produced them, — the tremendous deluges of former times, — need no longer excite the astonishment and wonder of the geologist as he contemplates the picture here suggested. So perfect an adaptation of means to the end accomplished, must, as he reflects upon it, tend to establish in his mind the truth of our theory, that every portion of the solid surface of the globe has been subjected, over and over again, to the action of the powerful aqueous forces consequent on sharp contrasts of seasons.

Professor Cazin gives, in the paragraph following, a graphic summary of Professor Tyndall's views of "the glacial epoch." He says, —

"Will the supposition of a transient cooling, a diminution of the solar action, or the passage of the earth through excessively cold regions of the heavens, explain this epoch? Tyndall has made a remark which has happily elucidated this question. Glaciers are the condensers of the ocean: to allow a large accumulation of ice on the mountains, the evaporation from the surface of the seas must be considerable, and therefore the sun must furnish more, rather than less, heat. To suppose that the glaciers augment in consequence of the suppression of the solar heat, is the same as trying to increase the distilling powers of a distillatory apparatus by diminishing the fire under the boiler. Solar heat could not, therefore, have been less active in the glacial epoch than at present. 'We have only to suppose,' says Tyndall, 'a more powerful condensing apparatus than at present. For this it would be sufficient that the mountains were higher during the glacial epoch than they are now;' and he goes on to say that Switzerland and other countries where traces of ancient glaciers occur, must, therefore, have "suffered a slow sinking, due to the gradual sinking of the earth's crust." *

Recognizing the truth of the assertion that the sun must have furnished more heat in the Glacial period, in middle and high latitudes than it does now, and assigning an adequate cause for the same, which others have failed to do, we cannot admit that the reason advanced in support of that part of the hypothesis making imaginary ranges of exceedingly lofty mountains the principal agency in condensing the superabundant vapor, is sufficiently conclusive.

In order to afford adequate support to the above view, it is necessary that all the mountains of the earth which exhibit the traces of more extensive systems of glaciation than those of the present (a classification embracing them all, even those of the tropics, as we shall soon see), must have been subjected to a contemporaneous elevation far above their present height. All must have retained the augmented altitude for the same length of time, and all must have sunk simultaneously to the original level. Not only must we imagine a general upheaval of the mountains, but we must also go so far as to admit that all parts of the earth's surface bearing marks of glacial action experienced, at the same time, a proportionate rise.

A very definite result of extensive modern observations is the establishment of the fact that upheaval and subsidence are proportional to each other. An elevation of the earth's crust at any given point involves a corresponding depression at some other, preserving thus not only the relative proportion of land and sea, but also the average height of land relative to the sea level. But if we choose to assume that only the mountains were elevated, and that the general level of the surrounding valleys and plains underwent a corresponding depression,

* The Phenomena and Laws of Heat, p. 262.

or even remained stationary, no considerable excess of glacial action could ensue; for if the sun was no hotter than now, the increased altitude of the mountain would carry the glaciers but little further into the plain; and under an augmentation of the solar heat they could hardly be expected to reach even their present terminal bounds. It is, therefore, plain that the large increase in the dimensions of the glaciers of the period in question could not have been the result of an increased elevation of the mountains, the assumption that such was the case being wholly unwarranted, and unsupported by any kind of proof whatever.

The solar heat, however, was more efficient at that time than it is now; at least in middle and high latitudes. This fact is shown by the increased evaporation necessary to sustain the excessive meteorological activity characterizing the Glacial period. But an increase of solar heat under the prevailing status, or present relative positions of the earth and sun, involves an increase in the whole annual mean of the terrestrial temperature. Such increase would, therefore, mitigate the cold of the winters in the same proportion as it would augment the heat of the summers. In this case there would be more evaporation and rain, but certainly there would be less ice and snow, an excess of which is implied in the distinctive title of the epoch. We are absolutely certain that there was *more* ice and snow than now, and therefore it must have been colder. But as the increase of heat could not have been absolute, so neither could have been the increased cold; for, as Tyndall observes, a decrease of solar heat would tend to arrest the distillatory process. There would be snow and ice, it is true, but in largely decreased quantity, consequent on the diminished evaporation of the water of the sea. But even should large masses of ice and snow accumulate, the reduced quantity of solar heat would be insufficient to effect the sudden liquefaction of the same, so as to produce the diluvial consequences characteristic of the period; and it is evident that the excess of atmospheric moisture would reach the sea in a manner analogous to that in which the present excess in polar climes does at the present time.

Both an absolute increase or an absolute decrease of the solar heat are thus equally inconsistent with the phenomena of the Glacial period. The comparison between the evaporation and precipitation of atmospheric moisture and the process of distillation is apt, for the analogy between the two is complete. It would be as absurd, therefore, to attribute the meteorological activity of Glacial times to either an increase or a diminution of the solar influence as it would to suppose that a chemist, conducting the distillatory process with the heat of the retort exactly graduated to the condensing power or cold of the receiver, could hasten the operation, on the one hand, by increasing the heat of the retort and of the receiver proportionately, or, on the other, by diminishing the heat of the retort, and effecting a proportionate cooling of the condenser. To succeed, it is manifest that he must raise the temperature of his retort, and reduce the temperature of his receiver. The increased evaporation in Glacial times shows that it was hotter than now, and the increased condensation and precipitation of atmospheric vapor show that it was also colder. The theory of thet ransverse revolution of the earth satisfactorily explains this seeming paradox, exhibiting to us, as it does, vast intervals of time in each rotation during which, in certain latitudes, or those in which ancient Glacial phenomena most abound, in some cases all, and in others nearly all, of the whole annual amount of heat received from the sun was concentrated upon the summers; the winters being, at the same time, determined by a more or less complete deprivation of the solar heat, as now seen in polar latitudes.

The phenomena of the Glacial period, therefore, coincide with and become possible only under sharp contrasts of seasons, such as are wholly inconsistent with the present status. Such contrasts can have proceeded only from changes in the relative position of the earth and sun consequent on increased obliquity of the ecliptic to the equator. Professor Tyndall's suggestion is therefore correct, that in the Glacial period, the heat of the sun (at least in middle and high latitude) was in excess of what it is now; but this excess was confined to the summer seasons, the cold of the winters being proportionally severe.

It must further be seen that even the meteorological phenomena of the winters of the present time, to say nothing of that pertaining to those of the Glacial period, would be wanting in an epoch of excessive cold, or a cosmic winter. The fact that a very low temperature arrests, to a great extent, the process of evaporation, and also prevents the precipitation of whatever moisture may exist in the atmosphere, — snow seldom falling with the thermometer much below the freezing point, — seems to indicate that the vast accumulations of pure ice, which have been described as characteristic of the Glacial period, are purely imaginative, and could never have existed, even admitting the truth of the cosmic winter theory. But while excessive continuous cold would have a tendency to arrest all meteorological processes, sharp contrasts of heat and cold — of summer and winter — would, on the other hand, have the opposite effect to produce them in excess. If we imagine the ocean and seas retaining into the winter a considerable portion of the heat acquired during the hot summer, they would, undoubtedly, supply the moisture requisite for the display of meteorological phenomena on the grandest scale upon the land. Immense deposits of ice and snow would ensue, that is, immense as compared with what are now seen, but not larger than would be, under ordinary circumstances, liquefied at ordinary levels long before the sun of the succeeding summer should have attained its greatest altitude. These remarks are intended to apply only to the surface ice and snow of each winter. The manner in which the permanent glacial deposits of the Ice period were accumulated will be indicated hereafter. It would thus appear that the annual recurrence of these spring floods, comparatively small as regards those that have been imagined, but excessive in contrast with those of our own times, continuing throughout the hundreds of thousands of years embraced in the Glacial epoch, reasonably and consistently account for the manifest great excess of fluviatile and diluvial action at that time.

We thus recognize, in this period, in their highest state of activity, those natural forces whose office it is to effect the periodical transformations to which the superficial world is continually subjected. Excessive heat and moisture in summer, and excessive cold and glaciation in winter, rending and breaking up, disintegrating and decomposing the rocky crust, are the agents supplying the material, and which at the same time, by glacial and diluvial means, also accomplish, on a comparatively large scale, each recurring season, the transportation and deposition of such material. This view, obviously, furnishes us with a theory of geological denudation and deposit much more consonant with observed facts than can be obtained by any possible effects of the breaking up of a single gigantic cosmic winter.

The uninterrupted course of life throughout this period is also conformable to and in strict harmony with the theory of a gradual change in the position of the earth with regard to the sun. It may, indeed, at first, be startling to reflect on the large amount of climatal change each and every portion of the earth's surface must undergo within the time required to accomplish each semi-transverse revolution; but we are constantly to bear in mind, first, that, possibly with the exception of circumpolar areas during the time of equatorial coincidence with the ecliptic, no climatal conditions follow such revolution in any part of the world, incompatible with the largest development of life; and secondly, that ample time elapses, while the climatal changes are in progress, to accomplish whatever amount of modification may be needed in the organic world to meet the ever-changing conditions of life. Some idea of the slow rate at which this motion of the earth, and the changes consequent upon it, — climatal, inorganic, and organic, — proceeds, may be acquired by supposing an intelligent being endowed with life and memory for a period of ten thousand years. Such an individual, living at any time during the semi-transverse revolution, either in its interval of change or of repose, would experience no sensible results of the movement; and throughout the whole course of his life might continue to be a firm believer in the "stability of nature," so far, at least, as his unaided senses would be able to instruct him to the contrary.

Superficial Changes of the Glacial Period. — It may be objected that the phenomena of the ice period could not have been consequent on sharp contrasts of seasons, — on winters of arctic severity alternating with summers of tropical warmth, — because the heats of each succeeding summer would effectually liquefy the deposits of snow and ice of the preceding winters; so that the vast glacial masses known to have been in existence at that time, would have been impossible under seasons of this description. This objection, plausible enough at first sight, loses its force when we consider the circumstances under which the ancient glaciers were accumulated. The prevailing ideas of the glaciers of the Ice period are derived from those now to be seen in high latitudes and lofty mountains; and, consequently, the former are supposed to have been, like the latter, masses of pure ice, but slightly intermixed with foreign matter. When we consider, however, the widely different conditions under which ancient and modern glaciation proceeded, it becomes at once apparent that no very close analogy is to be expected in the results. We, doubtless, may approximate the true theory of ancient glacial phenomena, by imagining the effect of polar winters upon the water-courses of lower latitudes.

It is well known that the terrestrial surface radiates heat with much greater facility than it absorbs it. There would be, therefore, an evident tendency towards a general reduction of the temperature of the earth's crust as the winters became more and more rigorous, notwithstanding the corresponding relative increase in the warmth of the summers. The temperature of the earth becoming thus largely reduced to a considerable depth, when the cold of winter should come to be determined by the entire absence of the sun, the frost would certainly penetrate the ground far deeper than it does now, solidifying not only the water of shallow lakes and rivers, but also the earth beneath their beds. It is evident, too, that with the sun's heat distributed over the earth, as it would be during a period of great obliquity, the cold of winter would occasionally be interrupted by brief intervals of warmth; for, even at the present time, arctic voyagers assure us that the thermometer is subject to extraordinary variations as far north as they have yet attained, the most intense cold giving way in the course of a few hours to a temperature sufficiently mild to admit of copious falls of rain. Thaws of a similar character occurring in the midst of one of the rigorous winters of the Glacial period, while they would have no effect upon the lower part, would largely augment, on the return of the cold, the upper surface of the frozen mass, while those of the succeeding spring, with their attendant floods, by bringing down and depositing upon the ice more or less of the loose material of neighboring high lands, would interpose an effectual obstacle, during the following summer, to the liquefaction of the frozen deposit. In this connection is to be taken into account the slight protection required to preserve, by artificial means, comparatively small quantities of ice through the heats of summer, and also the difficulty of effecting the liquefaction of ice through the application of heat to the upper surface alone. Even if the larger part of the ice formation of a single winter should melt during the subsequent summer, whatever portion of it might remain would augment that of the following winter, and the frost having once effected a permanent lodgment in the ground, the process above indicated continued for hundreds and perhaps thousands of centuries, would suffice to fill up at least the valleys of the rivers with ice deposits of various descriptions. A practical illustration of the condition of the earth above indicated is now to be found in high latitudes, where the mean annual temperature has never, since the Inter-Glacial period, been high enough to melt the subterranean glacial deposits, as has been the case in more southern localities.

"Humboldt remarks," says Lyell, "that near the mouths of the Lena a considerable thickness of frozen soil may be found at all seasons at the depth of a few feet. . . . According to Professor Von Baer of St. Petersburg, the ground is now permanently frozen to the depth of 400 feet at the town of Yakutzk, on the western bank of the Lena, in lat. 62° N., 600 miles distant from the Polar Sea. Mr. Hedenstrom tells us that, throughout a wide area in Siberia, the boundary cliffs of the lakes and rivers consist of alternate layers of earthy materials and ice, in horizontal stratification; and Mr. Middendorf told me in 1846, that, in his tour there three years before, he had bored in Siberia to the depth of seventy feet, and, after passing through much frozen soil mixed with ice, had come down upon a solid mass of pure transparent ice, the thickness of which, after penetrating two or three yards, they did not ascertain." *

The above passage furnishes a key to the satisfactory solution of the problems connected with the superficial states and changes of the earth during the Glacial period.

* Prin. of Geol., chap. x. p. 187.

It leads, in the first place, to a knowledge of the chief agency by means of which the river systems of entire continents of one geological period are changed to those of another, its successor; for a stream which, during the period of increasing obliquity, had been gradually raised by glacial and diluvial processes, as above described, to a considerable height above its former level, would be very unlikely, when the cold of the winters became ameliorated from decreasing obliquity, to resume its original channel.

It accounts, naturally and consistently, for the contorted and broken stratification of the alluvial deposits of the drift period. A certain proportion of the original formation of ice and alluvium, on the return of more equable seasons, would be washed out and transported elsewhere, but the part remaining in position, as the icy portion of it gradually thawed, would settle down and assume the appearance now observed. The layers of earthy material originally disposed in horizontal stratification would be bent, broken, and distorted; and in cases where the alternating ice was of considerable and irregular thickness, every trace of former stratification might be lost. This explanation appears much more reasonable than that based on the supposition that all of these deposits assumed their present order of arrangement through the agency of running water alone.

That the principles involved in our method of reducing to system the phenomena of the Glacial period are substantially correct, is further shown by its admitting of the general surface of the earth being free, during summer, from ice and snow throughout the whole interval of extreme obliquity, thus indicating a way in which all the forms of life, both vegetable and animal, were enabled to continue in existence under all its vicissitudes; and, contrasting such method with the hypothesis which assumes that are as continental in extent, during the Glacial period, were buried beneath superincumbent deposits of pure ice from six to ten thousand feet in thickness, — a theory sufficiently disproved by the continuity of organic life in that era, — we may with perfect confidence conclude that, in the whole history of the world, there has never been any approach to the conditions of a cosmic or geologic winter such as has been imagined and described; the phenomena giving rise to the idea being susceptible of a natural and reasonable solution, independent of such extreme and improbable means, under the theory of the transverse rotation of the earth.

Inter-Glacial Period. — Besides those traces of warm seasons discernible in the drift which we have considered as effects of the hot summers incident to great obliquity of the ecliptic to the equator, there is also evidence pointing to an interval of considerable length within the Glacial epoch when the severe cold of the winters became mitigated to some extent, with no corresponding diminution of the summer heat; and geologists have been led, accordingly, to entertain the opinion that a long intercalated period of warmth, distinct from the supposed lesser periods before referred to, characterized the middle portion of great cosmic winter. It has been denominated the "Inter-Glacial Period."

The reader will observe in the citation following, in which are described the phenomena indicative of this interval of milder weather, that such phenomena have been found only in Middle Europe, and seem more noticeable in the vicinity of the forty-fifth parallel of latitude than elsewhere; a fact that we shall find, as we proceed, to be of great significance. Under the simple and natural method of accounting for the Inter-Glacial period furnished by our system, it is to be referred to the time of coincidence of the terrestrial axis of diurnal rotation with the plane of the earth's orbit, or the ecliptic, and is the result of the peculiar seasons produced in the latitude above mentioned by such a position of the earth with regard to the sun. It is manifest that when the earth's axis is in this position, the whole annual aggregate of solar heat received at the earth will be distributed very nearly, if not quite equally over all latitudes, however unequally the share of any given latitude may be distributed throughout the seasons of the year. In the space between the equator and the poles every gradation of climate would be found, from the complicated arrangement involving two summers and two winters in each year at the former, and the simple single day and night constituting the summer and winter into which the year at the poles would be divided. It is at this point in the semi-transverse revolution of the earth that equatorial areas receive the smallest share they ever do of the annual aggregate of solar heat, while polar and middle latitudes receive the most. The winter at the poles would be determined, as it is now, by a six months' absence of the sun. At the forty-fifth parallel, however, the solar light and heat would be enjoyed for nine months, — one third of the same being continuous day, — and its absence would cover but the remaining three months of the year. The resulting climatal conditions may be supposed fully adequate to produce the phenomena of the Inter-Glacial period.

After premising that the excessive cold of the Glacial epoch "lasted for a long series of ages, although not always with the same intensity," Lyell illustrates the fact of such cold having been intermitted or sometimes mitigated for a season by mentioning what the late Hugh Miller called 'striated pavements.' "These," he continues, "consist of horizontal surfaces of boulder clay, in which the imbedded boulders are seen to have been subjected to a process of abrasion similar to that which the solid rock below had previously undergone. In such instances large stones or blocks fixed in the clay have not only their original and independent striæ, but have subsequently suffered a new striation which is parallel and persistent across them all. . . . Some examples of this second striation may have been due to the friction of icebergs on the bed of the sea during a period of submergence; others to a second advance of land glaciers over moraines of older date.

"M. Morlot and others have adduced abundant evidence of two glacial periods in the Alps, during the first of which the glaciers attained colossal dimensions, filling the great valley of Switzerland with ice, which reached from the Alps to the Jura, while on the southern side of the great chain other contemporaneous glaciers invaded the plains of the Po, where they have left moraines of truly gigantic dimensions. After these huge glaciers had retreated for a time, they advanced again, and though not on so large a scale, they still vastly exceeded in size the largest Swiss glaciers of our day. The interval of milder weather, marked by the decrease of snow and ice in the Alps, has been called by Professor Heer the Inter-Glacial period, which must have been of considerable duration, for it gave time for the accumulation of dense beds of lignite, like those at Dürnten, and other localities near Zürich. During this intercalated series of warmer seasons the climate is supposed by Heer to have closely resembled that now experienced in Switzerland. He infers this from the fossil flora of the lignite, especially from the occurrence of cones of the Scotch and spruce firs, and the leaves of the oak and yew, all of living species, as well as from the seeds of certain marsh plants. The insects also, and the fresh-water shells, tell the same tale. Among the mammalia occurring in the lignite-bearing shales of Dürnten are an elephant (*E. antiquus*), an extinct species of bear (*Ursa spelæus*), and a rhinoceros different from *R. tichorhinus*. That the formation of the shale and lignite containing the above-mentioned remains were both preceded and followed by periods of greater cold is shown on the one hand by the polished and striated rock surfaces on which the shale and lignite rest, and on the other by the large size of the erratic blocks which are superimposed upon them.

"In England the lignite, or Forest Bed as it is called, of Cromer, on the Norfolk coast, presents a singular analogy to that of Dürnten above described. It contains in like manner the cones of the spruce and the Scotch fir, and the seeds and leaves of marsh plants, and some shells and mammalia in common with the Swiss deposit. It was also preceded and followed by a period of greater cold. The antecedence of a colder climate is proved by the arctic character of a large proportion of the shells of living species included in the marine strata of Chillesford, near Ipswich, in lat. 52° N., which, according to the observations of Messrs. Prestwich and Searles Wood, are more ancient than the forest or lignite bed. On the other hand, that the Forest Bed of Cromer was followed by an era of severe cold, is shown by the fact that it underlies the great mass of glacial drift, which is in part unstratified, and contains boulders and angular blocks transported from great distances, and some of them exhibiting polished and striated surfaces." *

* Prin. of Geol., vol. i. p. 194.

In order to facilitate the reduction of the phenomena of the Inter-Glacial period, — for our present purpose sufficiently described above — to the new system, we shall, in our retrospect, anticipate so much as will be required to allow us to approach it in a direction opposite to our usual course. According to the prevailing geological system, the next formation antecedent to the drift is the Pliocene, the strata of which, according to our authority, indicate a gradual transition from a warmer to a colder climate. "When we pass beyond the ages when a colder temperature prevailed, and, receding a step further into the past, examine the fossils of the British Pliocene strata, we find in the earliest or lowest members of them very interesting proofs of a climate warmer than that now prevailing in England, and more resembling that of the Mediterranean. As we ascend in the series, the shells of successive groups of strata, provincially called crag in Norfolk and Suffolk, are seen to consist less and less of southern species, while the number of northern form is always augmenting, until in the uppermost or newest groups, in which almost all the shells are of living species, the fauna is very arctic in character, and that even in the 52nd and 54th degrees of North latitude." *

This appearance of slowly decreasing warmth extending throughout the whole of the so-called Pliocene period, is allowed by Darwin and others to be the coming on of the Glacial period, and is clearly due to a gradual increase in the inclination from perpendicularity with the ecliptic of the earth's axis of rotation, just as the subsequent appearance of a gradually returning warmth is consequent upon the gradual decrease of such inclination.

The impression that a warmer climate than is now experienced prevailed at the time the lower members of Pliocene group were deposited, founded on comparisons of the organisms of the two eras, when according to our theory the inclination must have been approximately the same, has arisen, no doubt, from the fact that the organic forms of the present time, being removed by only a comparatively brief interval from the Glacial period, still retain more of the arctic characteristics then impressed upon

* Prin. of Geol., vol. i. p. 197.

them than the others; separated as these last were from the preceding Glacial epoch by an interval of time five or six times greater.

Commencing, then, at that point in the present or Pliocenic semi-transverse revolution, when with reversal of the poles, the inclination of the earth's axis was the same as now, and when, in consequence, climate was essentially the same, — a point dating back almost precisely a million of years ago, — to follow increasing inclination as the cause of the before-mentioned apparant diminution of the terrestrial temperature, and more particularly its effect upon the climate of places situated in the forty-fifth parallels of latitude, we find that by gradually *increasing* the altitude of the sun and the length of the days in summer and proportionally *decreasing* the altitude of the sun and length of the days in winter, sharper and sharper contrasts of summer and winter are constantly produced, until the time arrives when the longest day in summer and the longest night in winter continue each throughout the whole twenty-four hours. During the long day, the sun would appear to revolve about the celestial pole at an angular distance of 45° from it, touching the zenith at noon, and impinging upon the horizon at midnight. The long night of midwinter would be determined by a similar revolution of the sun about the opposite celestial pole, its whole course, however, being below the horizon, except at a single point where it would touch the same; determining, at the same time, the long day in the opposite hemisphere. These being the effects, in latitude 45°, of an inclination of 45° in the earth's axis, it is at this stage of the transverse revolution that we locate the commencement of the Glacial period. It must be confessed, however, that considerable research and calculation will be required to determine the exact climatic results in all the different latitudes that must be entailed by the various positions assumed by the earth with regard to the sun during a half revolution, and also at what point in such revolution the appearance of greatest cold would be produced. It is, indeed, not only possible, but extremely probable, that the most marked traces of extreme continuous cold which occur in middle and high latitudes are to be attributed rather to those polar glacial periods incident to the coincidence of the ecliptic with the equator, than to those produced by its perpendicularity to the same, like the one we are now considering.

Assuming that in latitude 45° the most pronounced glacial phenomena would be produced by an inclination of from 45° to 67½°, as the motion continues and the earth's axis approaches coincidence with the ecliptic, the long day and the long night would each increase in length, until at the time of such coincidence they would attain a duration of three months respectively. At the commencement of the long day the sun would appear to revolve about that point in the heavens constituting the celestial pole, touching the zenith at noon, and the horizon at midnight, as before described; thence in gradually contracting circles, approaching the celestial pole, reaching it at midsummer, the descent being accomplished in reverse order. A similar movement of the sun about the opposite celestial pole, below the horizon, would at the latitude in question determine the winter. The entire year in latitude 45°, would, therefore, consist, first, of an interval of three months, exhibiting alternation of day and night; second, a single continuous day of three months' duration; third, another equal interval of days and nights; and fourthly, a night of three months, constituting the winter. This peculiar state of the seasons arises from the fact that at this time the plane of the sun's apparent path in the heavens is, at the equator, perpendicular to the horizon, inclining more and more to the same as the distance from the equator increases, becoming coincident with it at the poles. An inclination of forty-five degrees, in latitude 45°, would produce the variety of seasons above described; and it is to the length of the summers, and the unintermitted intensity of the solar influence during a large proportion of the time, and also to the shortness of the winters, tempered by the more equal distribution of the sun's heat over the earth, to which may properly be ascribed the phenomena of the Inter-Glacial period.

That the earlier portion of the Glacial epoch, or that part of it which preceded the interval of milder weather, was colder, and its glaciers more extensive than that which followed it, seems to be satisfactorily accounted for by the fact that the cold of the anterior portion must have been intensified by the vast deposits of ice and snow which had accumulated about the poles in previous ages, such accumulations evidently tending to materially reduce the mean temperature of middle latitudes. But, during the Inter-Glacial period, or while the earth's axis was coincident, or nearly so, with the ecliptic, when the annual amount of solar heat was equally distributed throughout all parts of the earth, those accumulations would disappear, and the climate of countries situated in middle latitudes would obviously show, during the later part of the Glacial epoch, the effect of this absence of polar ice-masses in a considerable augmentation of mean temperature over that of the earlier portion.

A general survey of that part of the geological record which extends from the commencement of the Pliocene period down to the present time, shows, at first, for a sufficient reason, a degree of terrestrial warmth somewhat in excess, *apparently*, of that now prevailing. To this climatic state succeeds the appearance of an extremely gradual and uniform decrease of temperature until we approach the middle of the Glacial epoch, at which time intervenes, in middle latitudes, the phenomena of the Inter-Glacial period. After this again ensues the normal cold of the epoch, somewhat mitigated in middle latitudes from the former excess by the absence of the polar masses of snow and ice melted during the Inter-Glacial period, and, thenceforward, a like gradual increase in temperature through all the subdivisions of Post-Tertiary times down to the modern age; the climatic conditions of the world regaining at this time substantially the original status. It seems quite unnecessary to add that these transitions, requiring, under any method of computation, a million of years for their consummation, cannot have been the effect of the combined action of any number of minor local agencies; for their consequences are identical and contemporaneously universal over all the world.

In this connection it is a curious and suggestive fact, that the traditions of ancient Eastern nations point to a time when the earth's axis of diurnal rotation was coincident with the orbital plane. Dr. Rees tells us, on the authority of Herodotus, that a tradition was current among the ancient Egyptians that the ecliptic was once perpendicular to the equator; and, according to Lyell, in the sacred volume of the Hindoos, called the "Ordinances of Menù," "It is declared that, at the North Pole, the year was divided into a long day and night, and that their long day was the northern, and their night the southern course of the sun." * These wonderful and interesting facts do not, however, oblige us to infer that the ancients were possessed of a complete theory of the transverse rotation of the earth; for it is hardly possible that such a system could have been deduced from purely astronomical data alone, without the aid of palæontological geology — a science undoubtedly of recent origin. If, as is more probably the case, the idea that the earth had formerly assumed such a position was derived from a knowledge of the apparent movement of the celestial pole, and that such movement was the result of a real motion of the earth, with a conjectural belief in its continuous or rotatory nature, it even then indicates an accuracy of observation and an advanced state of astronomical knowledge among those remote nations of antiquity many have supposed confined to modern times.

* Prin. of Geol., vol. i. p. 8.

CHAPTER III.

EVIDENCE SHOWING THE PREVALENCE OR DIVERSE CLIMATAL CONDITIONS IN FORMER TIMES. — LYELL'S RETROSPECT, CONCLUDED.

CLIMATE OF MIOCENE PERIOD. — POLAR WARMTH OF MIOCENE PERIOD. — MIOCENE ICE INTERVAL. — EOCENE CLIMATE. — EOCENE GLACIAL PERIOD. — CRETACEOUS PERIOD. — SIGNS OF ICE ACTION IN THE CRETACEOUS ERA. — THE GLACIAL INTERVALS OF OOLITIC, LIASSIC, AND TRIASSIC PERIODS, INDICATED BY THE EXTREME NORTHERLY RANGE OF SAURIAN TYPES. — SIGNS OF ICE ACTION IN TRIASSIC CONGLOMERATE. — PERMIAN EPOCH. — PERMIAN ICE INTERVAL. — CARBONIFEROUS PERIOD. — ICE INTERVAL OF CARBONIFEROUS PERIOD IMPLIED IN THE NORTHERLY RANGE OF ITS VEGETATION. — DEVONIAN PERIOD. — EVIDENCE OF ICE ACTION IN THE OLD RED SANDSTONE. — CONCLUDING REMARKS.

HAVING now endeavored to give a general idea of the synchronism, or correspondence in time, between the various phases of the Glacial period and the different positions the earth must have assumed with regard to the sun since the commencement of the Pliocene age, we cannot, in our survey of the monuments of older eras, expect to meet with evidence sufficiently detailed to indicate progress from one climatal state to another, as heretofore. We are now, from the present stand-point, to look back to an undetermined succession of glacial periods and their correlative intervals of repose, and we must not fail to bear continually in mind the fact that the superficial changes brought about by each recurring period of cold, eradicate to a great extent the work of its immediate predecessor. An observation of Lyell's relative to the danger of underestimating the quantity of time embraced in the "cold epoch" seems pertinent in this connection. He says, "In proportion as the ice increases in thickness, it cancels all marks of antecedent glaciation. The grinding action of the great ice-sheet which now envelops Greenland illustrates this process. Were that ice to melt, it would require as much skill to detect the evidence of the moraines and erratics of an older time as in the case of a palimpsest to recover the work of the original author, which had been purposely washed out to make room for the new manuscript." * Consequently it will be found, as we proceed, that, contrary to our experience thus far in our retrospect, in which the traces of cold have predominated, while evidences of warmth have been more exceptional, the older formations contain abundant proofs of seasons of warmth, while instances of facts tending to show the intercalation of seasons of cold are more rare, and, in some cases, entirely wanting. The more reliable and exact nature of evidence, founded on the direct action of natural forces upon inorganic matter, which, in the nature of things, must always exhibit the same effects, as contrasted with that based upon comparisons of ever-varying organic forms, may be assumed to offset the disparity in point of quantity between the two classes of testimony. It will, at any rate, be considered sufficient for our present purpose to generally find in each of the great geological periods fair presumptive evidence pointing to marked contrasts of climate. Thus, if certain members of any group of strata shall be found to contain proofs of the prevalence of a degree of warmth equal to or exceeding that of the present time, and certain other members of the same group shall, on the other hand, present unmistakable traces of glacial action, it will be held as sufficient evidence that such group, as a whole, represents in time the amount required to accomplish one half of a transverse revolution of the earth, or such part of the same as would produce a complete cycle of climatal change.

Before proceeding further in our retrospect, it may be well to again advert to the constant danger of error arising from our liability to look upon the world in its present state as the standard to which the worlds of all former eras are to be made to conform. We are to bear continually in mind the fact that it does not follow because the structure of certain species of animals and plants has remained *apparently* unchanged throughout the unintermittently varying conditions of life exhibited by a semi-transverse

* Prin. of Geol., vol. i. p. 196.

revolution of the earth, or its equivalent — a geological period — that their habits must necessarily have remained the same throughout such a vast lapse of time. The author upon whom we rely for our geological facts, although conscious of this source of error, seems frequently to forget that essentially tropical animals may, as in the case of the mammoth and its associates, become so modified as to endure winters of considerable severity, if such winters shall alternate with summers sufficiently tropical in aspect, although it must be conceded that no mitigation of the rigor of the winters could have compensated to them the absence of that essential attribute of the summers which they required for their development and sustenance. Thus it will be found in the quotations following, that the presence of certain classes of organisms is considered as implying a climate warmer than that now prevailing in the same locality, when in truth the fact may only indicate greater contrasts of seasons, and warmer summers.

In the marine formations of the Upper Miocene period, the era next preceding the Pliocene, we find, according to Lyell, that "a third or more of the testacea belong to living species, not a few of which are now inhabitants of more southern latitudes, and of the associated fossil species unknown as living, some belong to genera now characteristic of more southern climates. Although in Great Britain Upper Miocene strata are entirely wanting, they occur in Belgium and North Germany, where they contain shells of the genera conus, cancellaria, and oliva — forms all of them foreign to our seas as well as to our British Pliocene deposits, and proper to and indicative of a higher temperature.

"The French strata of the same age, called the Faluns of the Loire, point to similar inferences, and, like the contemporaneous beds of the Vienna basin, contain some fossil shells of species now living in Senegal, or off the western coast of Africa. The Upper Miocene flora and fauna of the whole of Central Europe afford unmistakable evidence of a climate approaching that now only experienced in subtropical regions. In one of the newest deposits of this Upper Miocene formation, Professor Heer has detected, at Œninghen, in Switzerland, the leaves, fruits, and sometimes flowers of about five hundred species of plants, in which we find a near resemblance to the flora of the Carolinas and other Southern States of the American Union. After selecting four hundred and eighty-three of these species as capable of comparison, specifically or generically, with plants now living, he finds that one hundred and thirty-one are such as might be referred to the temperate zone, two hundred and sixty-six to a sub-tropical, and eighty-five to a tropical latitude. In the present state of the globe, the Island of Madeira presents the nearest approach to such a flora. The proportion of arborescent as compared to the herbaceous plants is very great, and among the former the predominance of evergreens implies an absence of severe winter cold. A rich insect fauna, such as belongs to a warm climate, is also attested by the great number of the species of those genera which are most easily preservable in a fossil state. The reptiles which play so insignificant a part in the Pliocene fauna of Central and Northern Europe form a more conspicuous feature in these Miocene formations. At Œninghen there are two tortoises and three species of salamanders, one of them more gigantic in size than the living species of Japan. Bones of the monkey tribe are also met with in Upper Miocene strata near the foot of the Pyrenees in France. Among them is a gibbon, or long-armed ape, equal to man in stature, and the femur of a large species of this family has been detected by Dr. Kaup in strata of the same age at Eppelsheim, near Darmstadt, in a latitude which corresponds to the southern part of Cornwall. In Greece also, near Athens, the remains of Upper Miocene quadrumana have been met with, confirming the inferences as to the warm temperature of Europe previously drawn by naturalists from the fossils, shells, and corals of Touraine, Bordeaux, and Vienna." *

From the investigations of Dr. Falconer and Sir Proby Cautley, who collected, in 1837, a large number of fossil remains from the Siwâlik hills, which skirt the southern base of the Himalaya to the west of the river Jumna, it is inferred that at this time when the climate of Europe is supposed to have been sub-tropical, a still greater heat prevailed nearer the equator. "Here the abundance and variety of the fossil mammalia is prodigious, there being no less

* Prin. of Geol., vol. i. p. 198.

than seven species of proboscidians of the genera mastodon and elephant. With these a huge extinct four-horned ruminant, called Sivatherium, was found, as well as a camel, a hippopotamus, a hyena, and more than one species of monkey. The associated reptiles also bear witness to a temperature higher than that of any European strata of the same date; for, besides some extinct saurians larger than any now existing, we find among them the living crocodile of the Ganges, *C. biporcatus*, and the living gavial of the same river, besides a colossal extinct tortoise, of which the shell was no less than eight feet in diameter." *

In treating of Upper Miocene strata of the West Indies, he continues, —

"If again we turn to the Upper Miocene formations of the West Indies, those, for example, of Antigua, San Domingo, and Jamaica, we discover in them species of corals similar to those found in beds of the same age at Vienna, Bordeaux, and Turin, and some of which, as Dr. Duncan has shown (1863), have a near affinity to species now living in the Pacific (South Sea), Indian Ocean, and Red Sea. *They lead irresistibly to the opinion that there was a much greater analogy in those ages than there is now between the temperature of the West Indies in lat.* 18° *N. and that of Europe in lat.* 48° *N.*" †

The remarkable confirmation of the truth of the principle upon which our system is founded, contained in the foregoing paragraph, the concluding sentence of which I have italicized, can hardly fail of being noted by the most cursory reader. Every student who has mastered the rudiments of the physical sciences should be able to perceive that no more "analogy" could ever obtain between the climate of places separated by thirty degrees of latitude than does now, under the same inclination of the earth's axis. While the solar influence exhibits the same degree of potency it now does, and the terrestrial axis maintains its present position, it is absolutely impossible that there can be any essential departure from the generally prevailing climatic characteristics of the different latitudes, and a more palpable absurdity can hardly be suggested than that, *under these circumstances*, the climate

* Prin. of Geol., vol. i. p. 199. † Ibid, 200.

of the West Indian Islands and that of Europe could have assumed any greater degree of general resemblance to each other in former times than they do at present. The evidence, however, points "irresistibly" to the fact that such resemblance, or analogy, existed in Miocene times. The circumstances, therefore, must have been different.

If we ascribe this similarity of climate which evidently subsisted between places separated by thirty degrees of latitude, to the more equal distribution of the solar influence over the earth during the Inter-Glacial interval of the Miocene semi-transverse revolution, the problem is rationally and completely solved; for every latitude would then, during some part of the year, enjoy the effects of a vertical sun; although it is true that none would escape winters of more or less severity.

Admitting the geographical changes from the present status, which Lyell suggests as the cause of the analogy in Miocene times between the climates of Europe and the West Indies, to have actually existed, such cause being embraced in Dr. Duncan's conclusion,—"Not only that there was no Isthmus of Panama, but also that there was no great barrier of land or Atlantic continent separating the Miocene seas of Europe from the contemporaneous seas of the West Indies,"—there is not the slightest reason for supposing that those climatal effects would follow; for such an hypothesis necessitates the belief that, were the Isthmus of Panama now to subside beneath the waters of the sea, a temperate climate would thereby be entailed upon the West Indian Islands; and also that, were the "great barrier of land or Atlantic continent" removed, a tropical climate would ensue in countries removed by forty-eight degrees of latitude from the equator. It may safely be asserted that there are existing islands within eighteen or twenty degrees of the equator in situations fully as favorable for the development of a temperate climate as would be those of the West Indies, were the Isthmus of Panama removed, and communication established between the two oceans. There are also areas in latitudes 45° and 50°, where a tropical climate is quite as liable to prevail as in the countries above mentioned, in the absence of the great land barrier. But navigators would search in vain for anything analogous to a temperate climate within eighteen degrees of the equator, or anything resembling a tropical one forty-eight degrees from the same. Such climatic vagaries, are clearly impossible at the present time, under the present inclination of the earth's axis; and, if impossible now, they must have been equally so in any other former age of the world, under a like inclination.

According to our authority, Lower Miocene strata are found to contain the same indications of a warm, or even a still warmer climate than the Upper. "For," says he, "nearly all the genera of plants which in the Œninghen beds were mentioned as characteristic of temperate latitudes, are wanting in the Lower Miocene, while the tropical forms are more numerous. . . . About eighty other plants are enumerated by Heer, all of which would be cut off by such a winter as now prevails in Central and Southern Europe." *

As we recede further and further into the remote ages of the past, this species of evidence becomes less and less reliable. But few living forms of animals or plants are to be met with in Lower Miocene strata, and we still insist that it does not follow, because these few have transmitted unchanged their peculiar structure to their descendants of to-day, that these descendants have retained the peculiar habits of their progenitors throughout the millions of years, vicissitudes of climate, and changing conditions of life, which have intervened between their times. And if the element of doubt enters into the evidence afforded by these persistent types, does it not in a still greater degree qualify that derived from extinct species? While all may assent to the proposition that just deductions founded on comparisons either of extinct or persistent species approximate in a greater or less degree the truth, none will have the hardihood to insist on their entire accuracy. When, therefore, we are told that *tropical* forms are more numerous in the Lower than in the Upper members of the Miocene group, a valid objection seems to lie against the unqualified use of the term tropical. So-called extinct tropical species cannot be *structurally identical* with existing tropical species; and if not identical in structure, how can we, from their forms alone, estimate with precision the power

* Prin. of Geol., vol. i. p. 201.

they may have possessed of enduring cold? There is positively no reliable method, and, therefore, when we are told that eighty plants belonging to the Lower Miocene flora would be cut off by winters such as now prevail in the locations where their remains are found, the statement can hardly be deemed sufficiently exact to serve as a basis on which to found a theory of Miocene climate. There is no doubt but that the eighty species were essentially tropical plants, and it is equally true that if in habit they were the same as their nearest living analogues, they would have been cut off by the cold of an ordinary winter. The fact is, however, that they were modifications of what we term tropical species; and, as they continued to flourish, through whole geological periods, in the same localities, it is certain, whatever may be the true climatal theory, that while retaining the same form, they must have varied in habit sufficient to adapt themselves to the slowly but continually varying climatal conditions.

If we endeavor to arrive at an idea of the probable effect upon the vegetation of a tropical country of an insensibly graduated change from the prevailing climate to one that, while exhibiting the present summer temperature, should be characterized by winters such as are now experienced in "Central and Southern Europe," such change to proceed at so slow a rate as to require at least a half million of years for its accomplishment, it must seem exceedingly problematical to us whether such a change would effect either a very large amount of extinction or any extreme degree of structural modification in such flora. We cannot, of course, arrive at a knowledge of the precise effect, but we may be certain of this much, that it would be vastly different from that which would be produced by such a change brought about in a sudden and abrupt manner. Some species would undoubtedly become extinct, and all the rest would experience variations of habit which, in some instances, would induce structural change, though not in all; such variation being in some cases so marked as to amount, apparently, to the introduction of new or original species; while the total volume of vegetation would, on the whole, suffer but little, if any, diminution.

It may be worthy of remark that the assumed excess

of warmth of Miocene climate is predicated largely on evidence derived from marine organisms. Now, it is evident that when, from the more equal diffusion of the solar influence over the earth's surface at the time of coincidence of the earth's axis with the ecliptic, the polar ice-masses became liquefied, and no permanent surface accumulations anywhere remained except in mountainous localities, whatever vicissitudes of temperature might be experienced upon the land, the temperature of the waters of the ocean, from equator to the poles, would approximate a mean much nearer than they do now If tropical seas were somewhat cooler, those of temperate latitudes would be considerably warmer, and we may naturally suppose, even in the absence of proof, that marine tropical species would range even further northward than those of the land. No surprise, therefore, need be experienced under this explanation in finding Miocene shells of tropical type 45° to 50° from the equator, or that the climate of Middle and Southern Europe, during a portion of the Miocene period, bore a striking similarity to that now prevailing in the Island of Madeira.

Arctic Miocene Fossil Flora. — The evidence we have adduced in support of the theory of the transverse rotation of the earth, founded on the ascertained range of Pliocene animals and plants to the neighborhood of the pole, however conclusive it may be, is not more complete, to say the least, than the analogous testimony of similar phenomena accompanying the Miocene strata of arctic latitudes.

Although in the present instance, in consequence of the limits to which our inquiries are restricted, it may be necessary to assume the existence of an abundant Miocene arctic fauna, no valid objection can be brought against such an inference until extensive areas of the earth's surface can now be shown, covered with a profuse vegetation of a character like that hereafter indicated, and, at the same time, devoid of animated life. Wherever there is available food, there are always organisms to utilize it.

But if animals will flourish wherever they can obtain a continuous supply of proper food, readily adapting themselves to great varieties of climate, with plants there is another essential prerequisite. They require for their development, within certain limits, a determinate amount of the solar influence; that is to say, a tropical flora requires something analogous to a tropical climate, and a flora of the temperate zone something near the annual amount of warmth characterizing the latitudes of such zone.

Now, under the existing relative positions of the earth and sun, the climate of the several terrestrial zones must remain essentially the same; all other causes of variation in climate being comparatively trivial, and subordinate to that involved in the presence and absence of the sun, and the greater or less inclination of its rays. The annual amount of the solar influence anywhere experienced will determine the character of the vegetation there, and conversely, the character of the flora of any place will indicate with considerable accuracy the annual amount of the solar energy it experiences, although, perhaps, not so clearly, the mode of its distribution throughout the year. It would be just as impossible, therefore, for an abundant vegetation to flourish at the pole, under the existing degree of inclination of the earth's axis, as it would for the terrestrial day to ensue with the sun below the horizon, or the shades of night to prevail with it above the same.

Premising, then, that the volume and character of the vegetation of any locality is a correct index to the annual amount of the light and heat of the sun it enjoys, we will proceed with our collocation of facts.

"We find," says Lyell, "in certain beds of lignite or surturbrand in Iceland, recently examined by Professor Heer, an assemblage of fossil plants resembling in many respects that of Œninghen, before mentioned. Though not of so sub-tropical a character, they imply a warmth as much exceeding that now enjoyed in Iceland as did the temperature of the Upper Miocene flora of Central Europe surpass that of the vegetation now proper to the same region.

"The extent to which the Miocene flora flourished within the Arctic circle, even as far towards the pole as our exploring expeditions have penetrated, has been clearly pointed out by Professor Heer, in an important treatise on the fossil flora of the Arctic regions. In the numerous plates which illustrate this work, we see figures of more than sixty species of North Greenland fossil plants found opposite Disco Island, lat. 70° N. Among them are several species of Sequoia (*Wellingtonia*), with their male catkins and cones, agreeing specifically with Lower Miocene plants of Switzerland, Germany, or England. There are also seven other conifers, four poplars, two willows, three species of beech, four of oak (some of which have leaves half a foot long), a plane-tree, a walnut, a plum or prunus, a buckthorn, an andromeda, a daphnogene with large leathery leaves, and several other evergreens, some of extinct genera. The large-leaved trees imply, according to Heer, a high summer temperature, while the evergreens exclude the idea of a very cold winter. That these and other fossil plants from arctic localities really lived on the spot, and were not drifted thither by marine currents, is proved by the quantity of leaves pressed together, and in some cases associated with fruits, also by the marsh plants which accompany them, and by the upright trees with roots which were seen by Captain Inglefield and by Rink.

"Still further north in Spitzbergen, in lat. 78° 56′ N., no less than 95 species of plants are described by Heer, many of them agreeing specifically with North Greenland fossils. In this flora we observe Taxodium of two species, a hazel, poplar, alder, beech, plane-tree, lime (*Tilia*), and a potamogeton, which last indicates a fresh-water formation, accumulated on the spot. Such a vigorous growth of fossil trees, in a country within 12° of the pole, where there are now scarcely any shrubs except a dwarf willow, and where there are only a few herbaceous and cryptogamous plants, most of the surface being covered with snow and ice, is truly remarkable. When the fossils are compared with the Miocene species of Central Europe and Italy, many of them are found to be the same, and it is clear that the climate was not only much warmer than now, but the temperature of Europe and the Arctic circle was much less contrasted; nevertheless, the flora of Spitzbergen was by no means so sub-tropical at the era alluded to as was that of Switzerland, Germany, and Devonshire, for in the Lower Miocene period the difference of latitude made itself felt as now, although in a less degree. Professor Heer infers, with great probability, that pines, alders, poplars, willows, and other hardy genera reached the pole itself in Miocene times, if there was land there, because they range at present from 4 to 10 degrees farther north than the Taxodium, beech, plane and lime, which accompany them in a fossil state in the same formation at Spitzbergen. Some of the last-mentioned genera are in a higher latitude in Spitzbergen, by 8, 17, and 23 degrees, than the living representatives of the same genera. We cannot hesitate, therefore, to conclude that in Miocene times, when this vegetation flourished in Spitzbergen, North Greenland, and on the Mackenzie river, as well as Banks Land, and other circumpolar countries, there was no snow in the arctic regions, except on the summit of high mountains, and even there perhaps not lasting throughout the year." *

Excepting the reiterated assertion that the presence of evergreens necessarily implies very mild winters, and the conclusion that in Miocene times, when the above described vegetation flourished in arctic regions, there was no snow there except on the summits of high mountains, it would almost seem as if the facts and accompanying inferences contained in the above citation had been collated and arranged with special reference to the support of the theory of the transverse rotation of the earth. Even these exceptions are not inconsistent therewith, save when we accord an extreme significance to the terms employed. If winters are to be considered as not very cold unless rigorous enough to produce fatal effects upon the living evergreen plants of temperate latitudes, evidently those which prevailed in circumpolar regions during the Inter-Glacial period, free from the refrigerating effect of permanent ice masses like those of the present time, and ameliorated to a very great extent by warmth derived from the ocean, which would constitute a vast reservoir of heat, stored up during the intervening summers, such winters may very readily be conceived to have been not "very cold." So also it may be a matter of doubt whether there would be much if any snow, at least in the vicinity of the sea, whatever quantity might be precipitated annually upon inland countries.

But whatever may have been the character of the winters, of which we only know that they were not

* Prin. of Geol., vol. i. pp. 201–203.

inconsistent with the vigorous growth of a highly varied vegetation, such a flourishing vegetation unmistakably indicates an annual mean of temperature largely in excess of that now prevailing there; while, according to the highest authority, the nature of a conspicuous portion of the flora implies a high summer temperature. These, therefore, are established facts. At the time of deposition of these arctic strata, the summers were hot, the winters, at least, not excessively severe, and the whole annual mean of temperature highly favorable to a profuse vegetable growth. Well may one holding the opinion that the order of nature, as we now see it, has prevailed from the beginning, and must in future continue essentially unchanged, on beholding the fact of these phenomena, exclaim that it is *truly remarkable.* It is more than that, however. It is absolutely impossible; for, to produce these effects under the existing inclination of the terrestrial axis of rotation, an enormous increase in the potency of the solar influence would be required. An increase of 50° F. in the mean temperature at the poles would scarcely be adequate to induce this exuberant vegetable growth. To effect a rise of 1° at the pole, eight times that amount would be required at the equator. To raise the temperature at the pole 50°, under the present status, an increase of the sun's heat would therefore be necessary, sufficient to raise the temperature at the equator 400° above the present mean, the proportionate augmentation of heat in lat. 45° being about one half that amount. It is needless to add that such an exhibition of solar energy is largely in excess of the amount required to annihilate all life, vegetable and animal, upon the earth, excepting, perhaps, in arctic and antarctic regions. It is submitted, that these are no groundless hypothetical vagaries of the imagination, but are the inevitable natural results to which we must arrive while we persist in the futile endeavor to reconcile the above facts with the existing status.

The universality of the profuse Miocene fossil flora of arctic latitudes, the same occurring in widely separated places throughout the polar regions, is a sufficient refutation of any hypothesis that would seek to account for the phenomena through the agency of inferior local causes. Allowing that the chance combination of a number of those minor influences which have been suggested (to which attention will be more particularly directed hereafter) produced, in Spitzbergen, the climatal conditions upon which such flora was consequent, it is wholly out of the range of probability that a similar combination should, contemporaneously, produce identical consequences at Banks Land, on the Mackenzie River, North Greenland, Iceland, and, in short, all points within the arctic circle where the formation in question occurs and has been critically examined. The universality of the effect demonstrates that its cause, whatever that cause may have been, was also universal.

The same analogy that we have before observed to subsist between the climate of the West India Islands and that of Middle Europe, we also find to prevail at the same time between that of middle latitudes and of the poles: another proof of the dominant universality of the producing cause of these states and vicissitudes of climate. This climatal analogy was due to the more equal diffusion of the solar influence throughout all latitudes, consequent on the position of the earth with regard to the sun; and the difference of latitude which "made itself felt as now, although in a less degree," may be attributable less to the varying length and severity of the winters in the various latitudes, than to the primordial difference in the floras peculiar to those different latitudes.

Glacial Interval of the Miocene Period. — We have had occasion to designate the evidence of the reality of the transverse rotation of the earth adducible from the geological records of the Miocene epoch as being complete. Such evidence, however, cannot so be characterized unless it give some token, in middle latitudes, of glacial action similar to that incident to the Pliocene period. These tokens do exist; and Lyell, who confesses himself an unwilling witness, from personal observation, furnishes us with the most unequivocal testimony as to their authenticity. After adverting to the absence of organic evidence, which, for various reasons, can hardly be expected to exist at this time, in true glacial deposits of early date, he goes on to say: "But our geological records are far too fragmentary to entitle us positively to assume that, in the course of so vast a succession of ages, there were no oscillations of temperature analogous to those which certainly occurred between the close of the Newer Pliocene period and our own time. Professor Ramsay, who has so successfully devoted much time and thought to the search for indications of glacial action in remote eras, reminds us that a geologist must expect to encounter great difficulties in such investigations. If, at some future era, when large portions of the existing continents shall have been submerged and overspread with marine strata, and other parts of them destroyed by denudation, we should have the task assigned to us of detecting those spots where ancient land-surfaces had escaped destruction, or where erratic blocks and moraines of glaciers were extant, we might well despair of success. It rarely happens that we have opportunities of examining terrestrial surfaces of high antiquity, and when visible, their extent is always very limited. In the majority of cases they will consist of rocks incapable of receiving and preserving a glacial polish and striation. The least evanescent of the proofs of ice-action, which our era is likely to transmit to future ages, are, unquestionably, those large angular erratics which have been carried to great distances from their parent rocks; and wherever such masses occur in older strata they deserve particular attention. I shall proceed, therefore, to describe a formation of Miocene date, which I have myself examined, in which the position and size of the included blocks is such as to make it impossible at present to account for their transportation by any other cause than the buoyant power of ice.

"The marine deposits alluded to consist of strata of sandstone and conglomerate, and constitute a member of the Miocene formation of the Collina of Turin, a chain of hills in the suburbs of the capital of Piedmont, on the brow of which stands the church of the Superga. These strata have long been celebrated for containing a plentiful store of fossil shells of the same species as those of the faluns of Touraine, Bordeaux, and Vienna. The annexed diagram" (which from the importance of the subject we insert) "will give the reader some idea of the position of this conglomerate (*a*), which is highly inclined and conformable to the other strata which dip on each side to the north-west and south-east from the axis of the chain. I examined the district in 1857, in company with Signor Gastaldi, one of the ablest of the Italian geologists, and one well versed in glacial phenomena.

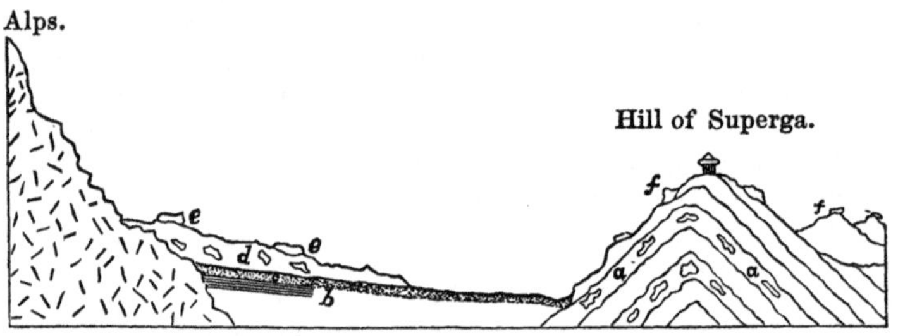

SECTION FROM THE ALPS TO THE HILL OF THE SUPERGA, SHOWING THE POSITION OF THE MIOCENE ERRATIC BLOCKS.

a. Conglomerates of Miocene age with large blocks.
b. Marine sub-Apennine or Pliocene strata.
c. Diluvium or ancient alluvium of various ages, some of it below the moraine *d.*
d. Moraine of Ivra of the Glacial period with erratic blocks.
e. Erratic blocks lying on the moraine *d.*
f. Miocene blocks washed out of the conglomerate *a*, and scattered over the hills of the Superga chain.

N. B. — The distance from the Alps to the Superga is about thirty miles.

"On this occasion I satisfied myself that Signor Gastaldi was right in supposing that the large blocks *f f*, lying on the surface of the hills, had been washed out of the beds *a a*, by the same action which has hollowed out the valleys. In other words, they have not been brought from a distance, as was once supposed, during the more modern or Post-Pliocene Glacial period, like the erratics *e*, which rest on the moraine *d*, but have been washed out of the Miocene beds in the immediate neighborhood, viz., the conglomerate *a*. This last is part of a regular series of strata, composed chiefly of sand of various degrees of coarseness, and of gravel, in which are rolled pebbles of greenstone (or diorite), limestone, porphyry, and some other rocks. Among them we occasionally meet with fragments of serpentine and greenstone, of enormous size, one of which I ascertained by measurement to be fourteen feet in its longest diameter. Signor Gastaldi has seen another in the same formation, twenty-six feet long; they are angular, and several of those which I saw exhibited some faint striæ and had one of their sides polished, in a manner much resembling that produced by glacial action. The whole thickness of the beds through which these blocks are dispersed varies from one hundred to one hundred and fifty feet. As yet they have yielded no organic remains, but they are covered by strata containing shells of the Upper Miocene formation, and they rest on Lower Miocene strata for the most part of fresh-water origin. The fauna and flora, both of the overlying and underlying rocks, have," (*apparently*, we may add), "the same subtropical character as those of Miocene date in Switzerland and in Central Europe generally. Hence the hypothesis of the transport of such large blocks by ice-action has naturally been resorted to most unwillingly, but in the present state of our knowledge it is the only one which appears tenable. The beds of sandstone alternating with those in which the blocks are enveloped exhibit no signs of having been tumultuously accumulated as by a flood. The erratics seem rather to have fallen quietly into their places. The nearest spots where any similar serpentine and greenstone occur are about twenty miles to the westward, but there has been so much subsidence of the country during the Miocene period, so much subsequent deposition of overlying miocene, pliocene, and alluvial deposits, and such changes in physical geography, that we cannot decide with any certainty as to the proximity or distance of the spots from which the blocks may have come." *

The author then proceeds to suggest methods of accounting for these, to him, anomalous phenomena of a glacial episode accompanying a formation intercalated in the very centre of the Miocene group of strata. As they occur, however, the most palpable difficulties are inseparable from them, and even the one at last settled upon as the "least objectionable hypothesis," viz., "a lofty mountain, with a glacier reaching the sea," seems hardly to afford him the satisfaction the earnest student of nature may be supposed to experience on first perceiving a nat-

* Prin. of Geol., vol. i. pp. 203–206.

ural and just relation to subsist between classes of well-ascertained facts and some favorite, but not fully determined hypothesis. The thickness of these beds of conglomerate, or rather of glacial drift material, which "in some parts of Piedmont is very great, far exceeding that seen in the vicinity of Turin," their great extent, and what may be assumed as their general original disposition previous to disruption by internal terrene forces, are facts strongly militating against the theory the author has judged the most probable, or, rather, the least objectionable. The same phenomena, however, in conjunction with the central position these deposits occupy in the Miocene series, — the absence in them of organic remains, — the massive erratic blocks disposed throughout them, one side of some of such blocks bearing the unmistakable traces of glacial striation and polish, altogether point with unerring certainty to the actual occurrence of a glacial interval analogous to the Pliocene during the Miocene geological period, such interval being consequent on the same adequate and natural cause as produced the glacial interval of our own geological period, viz., a large inclination of the terrestrial axis of diurnal rotation, and consequent vicissitudes of climate. They further indicate that this interval of cold winters occurred at its proper time in the Miocene cycle of climatal change, the peculiar phenomena incident to it occupying their appropriate place among the superficial consequences of that cycle upon the crust of the earth.

When we bear in mind the obvious tendency of a subsequent Ice period to cancel and annihilate the work of its predecessor, a tendency strongly insisted upon in the fore part of our last extended quotation, and vividly illustrated in a former one, the fact that such marked and tangible traces of a former glacial interval have escaped the destructive influences of that of the Pliocene age, may be considered as more than remarkable; it is truly wonderful.

On the whole, then, the inconsistency of the geological phenomena of the Miocene period with the present prevailing terrestrial status, and, on the other hand, their perfect harmony in every particular with the theory of the transverse rotation of the earth, is shown by the following considerations: —

1. The fossils of the Siwâlik hills; the deposits containing which are assumed merely to have been contemporary with the other Upper Miocene strata alluded to, indicate a higher mean temperature than now prevails in that part of Hindostan. Now, under the fundamental principle which underlies our system, that the position of the earth with regard to the sun primarily determines climate, any departure from the present status necessarily involves a corresponding change in the earth's relative position with regard to the sun. A concentration of the solar influence in the direction of the equator can be produced only by an approach of the ecliptic to coincidence with the equator. The facts, therefore, show that in the era of these deposits, the terrestrial axis of rotation, instead of exhibiting its present degree of inclination, was perpendicular, or nearly so, to the orbital plane.

2. The most indubitable evidences which exist of the intervention of a glacial interval in middle latitudes during the Miocene period are inconsistent alike with the idea of the prevalence, during such interval, of the terrestrial status indicated above, or that which prevails at the present time, and is explicable only under the theory of a great degree of inclination of the earth's axis.

3. The abundant overwhelming proofs of the coincidence of the earth's axis with the ecliptic, determining the Inter-Glacial interval of the Miocene epoch. These proofs are fouhd in the climatal analogy subsisting at that time between polar regions and Europe, between Europe and the West Indies, and, in view of evidence hereafter to be submitted, inferentially, between the West Indies and equatorial localities; all pointing to a general analogy of climate from equator to the poles; the wide extent of the deposits testifying to these peculiar climatal conditions, the effect of a more equal diffusion of the solar influence throughout all latitudes, showing, at least, its prevalence over all of the northern hemisphere; and last, but perhaps most decisive of all, the abundant vegetation that at the same time flourished in the immediate vicinity of the pole, which, in the nature of things, could only have been consequent upon such more equal diffusion of the solar energy.

We have therefore, in the geological record, competent

proof of three distinct positions of the earth's axis within the Miocene epoch, all different from that which it now occupies. In view of the imperfect manner in which Nature records her operations, can it not with truth be said that the evidence pointing to the Miocene semi-transverse revolution of the earth is complete? I cannot refrain from observing that if it does not so appear, it must be attributed rather to the inexperience and want of skill of the writer in arranging and presenting the facts and arguments bearing on the question, rather than to any paucity of such material. But as a just and upright judge and an intelligent and faithful jury ought never to permit a righteous cause to suffer through the inefficiency of its advocate, so it is to be hoped that the august tribunal before whom our theory must appear, will accord to it a fair and impartial trial upon its own intrinsic merits, without regard to any extraneous considerations whatever.

Eocene Climate. — The fossils and other monuments of the Eocene epoch, the next in our retrospect, as far as they go, indicate in the same manner as do those of the Miocene period, the same cycle of climatal changes; a cycle such as can have resulted only from a semi-transverse revolution of the earth. It would evidently be absurd to expect to find in each geological period as complete proofs of this motion as are to be met with in the Miocene or latest period that has accomplished the full half revolution. But if in the older geological epochs, taken collectively, we encounter repeated instances of these remarkable contrasts in the terrestrial climate, which in their nature can only have been the effects of different positions of the earth's axis, the fact cannot but be admitted as satisfactory corroborative testimony going to establish the truth of our theory. If each and every one of those epochs plainly shows, in various parts of the world, the consequences of several different positions of the terrestrial axis, the aggregate of proof to be derived from them all, taken together, is as conclusive if not more so than that derived from the Miocene period, although exhibiting, as the latter does, all the more pronounced climatal states resulting from such changes of position. Thus, although no mention is made of the discovery of organic remains indicative of an Eocene period of warmth at the pole, if the records of many of the other geological epochs establish the fact of the occurrence within them of such periods of arctic warmth, its absence in the Eocene can be attributed only to the imperfection of the geological record and our incomplete knowledge of the portion that remains entire.

It will be observed, as we recede from one remote epoch of antiquity to others still more remote, that organic evidence *seems* to show the prevalence of higher and higher degrees of temperature upon the surface of the earth. We claim, however, that there has not actually been, to any material extent, any such augmentation in the terrestrial temperature; but that the appearance is merely the result of an obvious tendency of every part of the earth's surface to transmit to later times more numerous and tangible organic traces of its eras of warmth than those of colder and, therefore, less productive seasons. It may also be not altogether a fanciful assumption that zoological, and, more especially, botanical genera have from their beginnings been approaching a closer and closer co-adaptation to the general climatic characteristics of particular zones; so that certain classes of forms which formerly were enabled to maintain existence over a large extent of the earth's surface, and under great variety of climate, may now have become by this means reduced to narrower limits.

For these reasons, in connection with others heretofore strongly insisted upon, the fossil fauna and flora of the Eocene period are erroneously supposed by Lyell to indicate an increase of temperature over that of the following, or Miocene age. "In the flora of the upper members of this great series," he says, "we find in the neighborhood of Paris and in the Isle of Wight, some plants which, like the palmetto, attest a warmer temperature. Among the accompanying reptiles, there are many crocodiles and tortoises, such as we now only meet with in more southern regions. In the Middle Eocene, as in the calcaire grossier, for example, near Paris, the marine testaceous fauna is richer and more varied than that now proper to seas so far north. The flora of the same division of the Tertiary period, as, for example, that of Alum Bay in the Isle of

Wight, of Monte Bolca in the North of Italy, or that of Aix en Provence in the South of France, comprises species and genera having a great affinity to Lower Miocene forms, but departing further than do these from the modern European type, and, according to Heer, resembling in many respects plants of the tropical regions of Australia and India.

"The nummulitic formation of this era is of world-wide extent, and contains many corals of large size, of genera now common in tropical seas, some of the same fossil species ranging from Scinde in India to the West Indies.

"If, lastly, we turn to the Lower Eocene strata, we find in the London clay of the Isle of Sheppey fossil fruits of the cocoa-nut, screw-pine, and custard-apple, reminding us of the hottest parts of the globe; and in the same beds are six species of *Nautilus*, and other genera of shells, such as Conus, Voluta, and Cancellaria, now only met with in warmer seas. The fish also of the same strata, of which fifty species have been described by Agassiz, are declared by him to be characteristic of hotter climates, and among the reptile are sea-snakes, crocodiles, and several species of turtle."*

Had only the fossilized osseous portions of the structure of the Wiljui rhinoceros and his Siberian compeers been transmitted to us from Glacial times, without the accompanying flesh, hair, wool, ice, and particles of half-masticated food, would they not with the same show of probability indicate the prevalence in their epoch of a purely tropical climate, as do the palmetto, cocoa-nut, screw-pine, and custard-apple of the Eocene period? In the absence of those direct, unequivocal proofs of cold, there would be absolutely no evidence whatever to rebut such, to all appearance, an extremely probable assumption. Now, if the progenitors of our large tropical mammals varied in habit from present types fifteen hundred centuries ago in a degree sufficient to enable them to endure rigorous winters, although it is very probable that plants require considerably more time than animals in which to effect their transformations, it must be far from impossible that those of tropical plants, removed from their descend-

* Prin. of Geol., vol. i. p. 207.

ants of to-day by an interval of more than fifty thousand centuries, or five millions of years, should be able to show an equal amount of differentiation.

Glacial Interval of the Eocene Period.—But whether Eocene climate was really warmer than that we now experience, or was only apparently so, such warmth or its appearance did not pervade the entire series. As in the Miocene, its traces occur in the upper and in the lower members, while its intermediate portion, like that of its successor, exhibits phenomena of a wholly different character. That the Eocene age was not without its Ice period is shown by what follows:—

"In a bed of coarse conglomerate of the Eocene period in the Alps, phenomena in many respects analogous to those of the neighborhood of Turin present themselves. This conglomerate is a subordinate member of that vast deposit of sandstone and shale which is provincially called 'flysch' and 'nagelflue,' and which, by its position (for it is devoid of organic remains), seems referable to the middle or 'nummulitic' portion of the great Eocene series. The well-known 'Vienna sandstone' is a member of this flysch, which extends for 300 miles at least, east and west, from Vienna to Switzerland, along the northern flanks of the Alps, and is again seen in the south, near Genoa, and in several parts of the Apennines, where it is called by the Italians 'macigno.' Its thickness is very great, amounting to several thousand feet, and occasionally, according to some authorities, to 6,000 feet. It is often finely stratified, and singularly barren of fossil remains, although in a few places it contains fucoids. Here and there, as in the Sihlthal, near the lake of Zürich, and in the Toggenburg in St. Gall, large blocks are enclosed in it, some of them angular and others rounded. These blocks are occasionally of limestone, and contain ammonites and other fossils of the oolitic and liassic formations, as described by Dr. Bachmann. Blocks also of a red variety of granite of a peculiar composition, not known *in situ* in any part of the Alps, occur in the same conglomerate of the flysch. In several places the blocks are 10 feet long, but at Habkeren, on the north side of the lake of Thun, many are seen of enormous dimensions, one of them being 105 feet in length, 90 in breadth, and 45 in height. They have lost their edges, either by friction or decomposition, but are not polished or striated.

"There has been a lively discussion as to whether the largest of the above-mentioned Habkeren blocks came out of the flysch, or were simply erratics of the Glacial period; but Escher von der Linth, Studer, Rütimeyer, and Bachmann are clearly of opinion that they have been washed out of the coarse conglomerate. The flysch of Bolgen, near Sonthofen, also contains foreign blocks of considerable size, and similar masses, as I am informed by Professor Suess, occur in Tertiary strata of the same age in the Carpathians and Apennines, but neither on them nor on any others have any glacial striæ been as yet observed. We have to account not only for the wonderful size of the granitic rocks, varying from 10 to 100 feet in diameter, but for the distance which they have travelled, which seems to be implied by our inability to refer them to any known source. They are distinguishable by their mineral character from all granitic erratics of the true or modern glacial period, such as are strewed over the surface of those districts of Switzerland where there is no outcrop of flysch conglomerate. The hypothesis that these huge masses were transported to their present sites by glaciers or floating ice, has been always objected to on the ground that the Eocene strata of nummulitic age in Switzerland, as well as in other parts of Europe, contain genera of fossil plants and animals characteristic of a warm climate. It has been particularly remarked by M. Desor, that the strata most nearly associated with the flysch in the Alps are rich in echinoderms of the Spatangus family, which have a decidedly tropical aspect. The entire absence of shells, or of organic remains generally, may perhaps be thought to favor a glacial origin for the flysch, but this negative character is too common in strata of every age to be of much value, except in connection with other proofs of intense cold. Nor must we disguise from ourselves the fact, that in the seas of polar regions where icebergs abound at present there is by no means any dearth of animal life. On the other hand, the regular stratification and even fine lamination of large portions of the flysch cannot be said to be inconsistent with a glacial origin, for on the Norfolk coast we see thinly laminated clays devoid of organic remains forming an integral part of unquestionable glacial deposits.

"The great thickness of the flysch, and the fucoids preserved in a few beds of it, lead to the conclusion that it was of marine origin. To imagine icebergs carrying such huge fragments of stone in so southern a latitude, and at a period immediately preceded and followed by the signs of a warm climate, is one of the most perplexing enigmas which the geologist has yet been called upon to solve." *

We need follow our author no further in his speculations on these phenomena, — the unmistakable traces of still another Glacial interval, — which under the assumption that the earth's axis has always maintained its present degree of inclination, are no more enigmatical and perplexing than are any other of the numerous instances of departure from the climatal status of to-day. It is the position of the earth with regard to the sun, involving various degrees of inclination of the solar rays, in different latitudes, that determines their respective climates, and nothing short of a change in that position can effect any material change in those climates. The occurrence therefore of a Glacial period in the latitude of Switzerland would be no greater anomaly than the prevalence of a genial climate at the poles under the present status; one would be just as impossible as the other. Both have repeatedly occurred, as we have already seen, and as we continue our investigations, the geological record will be found to exhibit most exact repetitions of the same round of testimony in the remaining periods of the world's history. There seems to be no possibility of mistaking the conclusion to which these facts legitimately tend.

The objection urged against the theory of an Eocene ice-period, that other deposits of "nummulitic age" contain fossils characteristic of a warm climate, is founded on the assumption that the flysch conglomerate and the other deposits alluded to were strictly contemporaneous. We are by no means sure, however, that such was the case. It is much more rational to suppose that the flysch conglomerate is referable to that portion of the Eocene semi-

* Prin. of Geol., vol. i. p. 207.

transverse revolution exhibiting an extreme degree of inclination of the terrestrial axis, and the other nummulitic strata containing fossils of "a decidedly tropical aspect" to the Inter-Glacial interval of the same period, than to suppose that a Glacial period could ensue in Switzerland at the same time that a tropical climate prevailed over all other parts of Europe, assuming the present status to have prevailed at that time.

We have, then, direct and accurate information of three distinct Glacial intervals or periods in the Tertiary age of the world. These intervals occur in regular sequence, and their phenomena are identical in character. Only that cause, therefore, which produced one of them can have produced the other; for nothing more absurd can be imagined than to suppose that regularly recurring cycles of similar events can result from chance combinations of dissimilar causes.

A partial resolution of all these climatal phenomena — seeming success in accounting for single instances of departure from present conditions — is not sufficient; the true system must and will show a natural and proper cause for each and every instance of such departure. The transverse rotation by constantly changing the position of the earth with regard to the sun, thereby continually varying the inclination of the sun's rays, and the proportional amount of the whole annual sum of the solar influence received in each degree of latitude, north or south of the equator, is the only conceivable agency that can have accomplished *all* the climatic variations which have transpired either in a single one, or in all of the geological epochs. This agency, and no other, *is* fully adequate to produce *all* the climatic vicissitudes that have ever occurred in all parts of the world; and these climatal vicissitudes are sufficient to effect all the superficial changes in the earth's crust, such as are involved in the processes of surface denudation and deposition of strata. We are not constrained, therefore, under our new system, to confine ourselves to one class of evidence, and that the least reliable of all, perhaps, to account for solitary instances of climatal variation, leaving all the others as unresolvable enigmas, entirely irreconcilable with the method employed in the apparently successful one, but any and every climatic status whatever, indicated in the geological record as having occurred at any time in the earth's history, and at any point upon its surface, is naturally and consistently referable to some statable, definite position of the earth with regard to the sun, incident to the transverse revolution. If this consideration is not of itself conclusive as to the truth of our theory, just and logical demonstration is of no value, and no processes of reasoning, however well fortified they may be by facts, can lead to any reliable conclusions whatever.

Climate of the Cretaceous Period. — Passing now to the eleventh chapter of the Principles, we find it prefaced by a brief summary of the conclusions to which the investigations concerning the climate of the world during Tertiary times are supposed to lead. The hitherto implied inference of a gradual decrease in the temperature of the earth from the earlier geological epochs down to the time of the Pliocene glacial interval, here finds definite expression. This opinion rests exclusively on organic evidence, which, as we have seen, is, at best, far from infallible, and must be extremely liable to mislead if too implicitly relied upon. The assertion that a miocene or eocene organic form which is identical in structure or approximately so with a living type, must therefore have possessed, throughout all intervening time, precisely the same habits, is erroneous, and has no weight, for such form must have experienced, under any theory of climate, great climatic changes, and, consequently, must have varied in habit to meet them. A species of animal maintaining existence from the equable climate of the Upper Miocene down through all the vicissitudes of the Glacial period to the comparative mildness of the present time, could not possibly, with the conditions of its life constantly changing, follow continually exactly the same course of life that it now does. Even in this geologically short period of time, the character of its subsistence may have changed many times, and it must have been hardened to various degrees of cold and sharp contrasts of seasons, and softened to various degrees of warmth and seasons of perpetual equinox. In cases where the changing conditions of life have induced modification of structure, we say that the species has become extinct, while persistent type are those endowed with the faculty of retaining their peculiar form through all the conditions incident to whole cycles of climatic change.

The summary manner in which the thrice repeated phenomena of excessive glaciation incident to the Tertiary age are disposed of, can hardly be allowed to pass without comment. In this connection our author says, "If, in certain localities in or near the Alps, some huge transported fragments of rock, enclosed in miocene and eocene conglomerates, seemed to require the aid of ice to bring them into the sites they now occupy, a local combination of geographical circumstances may perhaps be conceived, which might account for such exceptional cases without requiring a general refrigeration of climate at the times alluded to, or, still more probably, floating icebergs may, as suggested in explanation of the Habkeren erratics in the Alps, have brought large fragments from a great distance without requiring us to suppose a lower temperature than that now prevailing on the earth." *

How, we must inquire, do the geological facts alluded to constitute "the most perplexing enigmas which the geologist has yet been called upon to solve," if the above assumptions are to be regarded as affording a satisfactory solution of them? But they cannot be so regarded. It is, certainly, very easy to assert, in vague and indefinite terms, that a certain cause may be conceived capable, *perhaps*, of producing certain results. We may give loose rein to the imagination, if we choose, and conceive whatever vagaries we please; but in a case like this we ought to know the exact nature of these hypothetical fluctuations in physical geography, and also how such fluctuations could, according to the established laws of nature, have operated in the localities named so as to produce a temperature such as would admit of the formation of glaciers, or icebergs, capable of transporting to great distances blocks of granite containing nearly half a million cubic feet, at a time when all the rest of the world exhibited a tropical climate, with the exception, perhaps, of high polar latitudes, and even these enjoying a very large increase in warmth over that of the present time. It does indeed

* Prin. of Geol., vol. i. p. 212.

seem as if a proper course of reflection must convince any reasonable mind that no possible change in the relative quantity of sea and land in the vicinity, nor in the depth of the neighboring seas or elevation of the land, can, under the present status and present temperature of the globe, produce an arctic climate in Italy. If not now, even more impossible would it be under an augmentation of the terrestrial temperature. In either event, it would be as much out of the ordinary course of nature, as for a tropical climate to prevail at the poles, or an arctic one at the equator. It is indeed true that the agency named may modify the climate of limited areas to a very limited extent, but it is the mode in which the solar influence is exhibited at any point on the earth's surface that constitutes the dominant cause determining its climate; and it is absolutely impossible, with the earth's axis holding its present inclination, that anything else than the varieties of a torrid climate can prevail within the tropics, any but the varieties of a frigid climate within the polar circles, or other than the gradation of a temperate one within the intervening spaces. Any pronounced departure, therefore, in any latitude, from the generally prevailing climate of to-day, cannot but involve a change of status; and we may thus regard those huge Eocene and Miocene Alpine erratics, — *evidences of glaciation the most likely of any to be transmitted from one geological epoch to another,* — as the monuments of statedly recurring Ice periods, similar in every respect to the so-called Post-Pliocene Glacial period of later date.

It is assumed that these erratics of Alpine Miocene and Eocene conglomerates constitute phenomena of an exceptional character. If this does indeed seem to be the case at present, it is so, perhaps, because geological events have, in general, left a more legible record of themselves in Switzerland than they have in other countries, and, also, because the Alps have been more thoroughly explored by geologists than other mountains.

Resuming now our retrospect, and passing from the Tertiary to the Secondary formations "between which there are very few forms in common," we come to the Upper Cretaceous, or Chalk period. The evidence in regard to climate furnished by the fossils of this era, — the remains of organisms further removed from our time, and, consequently, more divergent, both in habit and structure from present forms, and, therefore, of greatly decreased reliability, — is held to point to a still further general augmentation of the terrestrial temperature, when, analogously to former instances cited, it can only properly be taken as indicating the climatic conditions of the time during the cretaceous semi-transverse revolution characterized either by the greatest annual mean of heat, or that portion of the same exhibiting the hottest summers. After endeavoring to derive support for the hypothesis of a warm climate by means of the fossil mollusca of the Chalk, the author continues as follows: —

"The plants of the Upper Cretaceous formation of Europe, so far as they are known, have such an affinity with the Eocene flora as to point in the same direction in regard to the existence of a high temperature. They contain a large number of dicotyledonous angiosperms, whereas the Lower Cretaceous rocks are characterized by the absence of these last, and by a predominance of cycads and of conifers of an araucarian type, and of ferns referred by some botanists to genera which also favor the hypothesis of a warm climate." * It will be observed as worthy of note, that botanists are not all agreed as to the bearing of the above facts. The degree of power living species of firs and ferns may be known to possess of enduring cold, cannot, by any means, be regarded as furnishing us with a reliable standard by which to estimate such degree of power in the extinct species of those genera of the Cretaceous period. In the millions of years constituting this period, and under the cycle of climatal states and changes incident to the same, it is quite probable that the cycads, conifers, and ferns alluded to, as well as the whole Cretaceous flora, during some portions of the cycle, may have exceeded living species, or nearest living allied forms of the same, in the power of enduring cold; while in other intervals of warm and equable temperature they may have become so modified as to be even more susceptible to its influence than kindred living types.

The Mesozoic or Secondary age of the world, from the general abundance of that class of forms, has been called the age of Reptiles; and the Cretaceous or latest period of

* Prin. of Geol., vol. i. p. 213.

that age is characterized by a large development, both in size and number, of these animals. This fact, taken in connection with the known habits of living species, is supposed to lead to the same conclusion of an enhanced temperature. In the uppermost member of the Cretaceous series, or Maestricht chalk, we find a "marked development of reptile life in regions where nothing analogous is now to be met with. Thus, in lat. 51° N., we encounter in St. Peter's Mount, Maestricht, the aquatic reptile called Mosasaurus, which was twenty-four feet in length. . . . The reader will observe, on consulting Owen's table of the distribution of reptiles in past geological ages, that of the five living orders, crocodiles, lizards, tortoises, snakes, and frogs, the two last-mentioned have not yet been traced as far back as the Secondary or Mesozoic periods, but the three first, the Crocodilia, Lacertia, and Chelonia, are met with in full strength in Cretaceous times, where they become associated with no less than three extinct orders, namely, Pterodactyles, Ichthyosaurs, and Plesiosaurs. Respecting the first of these, namely, the flying reptiles, it has been argued, that we have no right to assume that they required a hot climate, because they are so highly organized, and have so near an affinity to birds in structure, that they may have been warm-blooded, and as capable as birds of sustaining great cold. But the same argument will not apply to ichthyosaurs or to plesiosaurs, nor to the numerous chelonians which occur in the different divisions of the Cretaceous period, including the Wealden strata, in which large terrestrial saurians are so conspicuous." *

It is entirely useless to speculate on the physiological character and habits of extinct saurians, or other forms of life, which existed in an epoch so far removed from the present time as the Cretaceous. All we can know with certainty concerning them is, that they bore a greater or less resemblance, osteologically, to certain living forms. Here our knowledge ends, and here conjecture begins. Some of these anomalous forms — apparent compounds of reptile and mammal — seem to constitute important links in a graduated chain or series of modified types through which the large Tertiary mammals were developed from

* Prin. of Geol., vol. i. p. 213.

the fishes and reptiles of older periods. According to the doctrine of evolution, it is evident that warm-blooded animals must have been derived from cold-blooded progenitors. An aquatic animal becoming gradually amphibious in habit, and an amphibious one withdrawing by degrees from the water and becoming more and more terrestrial in nature by respiring larger and larger quantities of oxygen, would certainly have a tendency in this direction; and why may we not look upon these fossils as the remains of transitional types, intermediate between the two? Under this view, still less are we able to arrive at a definite knowledge of the conditions of life best adapted to their needs. If we cannot determine, with any degree of certainty, the character of the climate best suited to them, we certainly cannot know but what they were so constituted as to readily adapt themselves to each and every vicissitude of climate incident to the latitudes they frequented. Thus the Mosasaurus of Maestricht, instead of being, as Lyell supposes, a tropical or sub-tropical animal, may have been, as facts to the consideration of which we shall soon arrive seem to indicate, an inhabitant of waters in which floating ice was to be met with during certain seasons of the year, at least.

The difficulty of determining the climatal conditions to which an extinct animal was subject during its life, from an examination of the fossil remains of such animal, seems to be a constant source of anxiety to our author; and he is continually striving to render more stable the shaking ground beneath his feet. To that end, in discussing the question "how far extinct orders and genera may indicate temperature," * he says, —

"It has been objected, that in speculating on the habits and physiological constitution of plants and animals of an epoch so distant from our own as the Cretaceous, we enter a region of doubt and uncertainty, because even the Eocene species are distinct from the living ones, while the Cretaceous fossils differ as much from the Eocene as do the latter from living types. Dr. Fleming, therefore, when engaged in a controversy with Dean Conybeare, in 1830, as to the proofs of a hotter climate in the olden time, de-

* Prin. of Geol., vol. i. p. 214.

clared that the reasoning of his opponents was illogical, and their mode of dealing with the subject unfair. 'They were playing,' he said, 'with loaded dice;' for the large number of genera are now in tropical and sub-tropical zones, not because they could not live in colder regions, but simply because the land and sea in those zones are of wider extent, and support in equal areas a greater exuberance and variety of animal and vegetable forms. According, therefore, to the doctrine of chances, the majority of the genera of any past epoch, whether they be extinct or not, will have their nearest living analogues in hot countries. Many of them will be unrepresented in the colder parts of the globe, not because of their unsuitableness to the climate of such regions, but because of the comparative poverty of the fauna and flora of high latitudes. The fact, it is said, that the same genus has often species proper to the torrid, temperate, and frigid zones, is enough to demonstrate that it is on species alone that we can rely in questions of climate.

"The caution here enjoined," Lyell continues, "is by no means to be disregarded, but our scepticism on this head may be carried too far. If three assemblages of existing species were submitted to a good naturalist, one of them coming from arctic, another from temperate, and a third from tropical latitudes, he would be able at once to assign the quarter from which each of the three groups had been obtained, even though he might never have seen any one of the *species* before;" and he goes on to observe that this mode of reasoning enables us to arrive at conclusions respecting the temperature of periods when most of the genera, and many even of the orders of plants and animals, were different from those now living; thus greatly increasing our data of comparison when endeavoring to interpret the monuments of antecedent epochs, "since it is not merely to the living creation that we can appeal."

From our stand-point, however, there appears to be little danger of according too much weight to the reasoning of Dr. Fleming. It is true, indeed, that organic evidence is our main reliance when we attempt to determine the climatal state of the world in former periods; but, in its use, a source of error arises when we assign to it a degree of accuracy foreign to its nature. An assemblage of living organic forms will indicate to the naturalist, in a general manner, the climate of the locality from which it was obtained; that is to say, he will be able, on examining such an assemblage, to tell whether it was produced under an arctic, temperate, or torrid climate. This, however is as far as he will be able to go. He cannot estimate with any degree of exactitude, from such examination alone, the mean temperature of such locality, nor the manner or proportion in which the whole annual amount of solar heat there received may have been distributed between the different seasons of the year. If he cannot describe the present climatal conditions of any given region from a view of the organic forms now inhabiting it, still less closely will he approximate, when he endeavors to arrive at a knowledge of the prevailing climatic characteristics of remote epochs, by the mere inspection of the fossil remains of the unknown, extinct animals and plants which flourished in those epochs. From the beginning, all the individual species composing the organic world have been subject to a constant change, not only in habit, but in structure also, and this state of variation constitutes an unknown quantity, inseparable from the problem, for which we must not fail to make due allowance.

Glacial Interval of the Cretaceous Period. — The homogeneous character of the Upper Cretaceous or Chalk formation, and the general absence within the same of sand, pebbles, drift wood, and other signs of the vicinity of land, is accounted for by the fact that the deposition of the material entering into its composition took place at the bottom of deep seas. But in the chalk of the south-east of England, perfectly isolated single stones of considerable size are occasionally found, naturally exciting much surprise. In considering the question as to the manner in which such stones could have been carried so far out into the open sea, after adverting to a former attempt to assign their transportation to the agency of drift timber,— Darwin having observed stones as large as a man's head, which, entangled in the roots of floating trees, had, by them, been carried to long distances in mid-ocean, — the author admits, on reconsidering all the facts, that he must agree with Mr. Godwin-Austen, "that there are some cases which we cannot account for without introducing the agency of ice. Thus, for example, in 1857, there was found at Purley, near Croydon, in the body of white chalk, a group of stones, the largest of which consisted of syenite. This block had been broken up by the workmen before it was examined by any scientific observer, but the largest of the fragments was ascertained to be twelve inches in diameter in two directions, and to weigh upwards of twenty-four pounds. It was surrounded by granitic sand and pebbles of greenstone, and its dimensions rendered the hypothesis of transportation by drift timber inadmissible. There was, moreover, a total absence of carbonaceous matter, such as might have been looked for if a water-logged tree had sunk on the spot. Mr. Godwin-Austen, therefore, has suggested that the pebbles and loose sand must have been frozen into coast-ice, and then floated out to sea, and the stones, he observes, mineralogically considered, present just such an assemblage as might now be found on a beach on the coast of Norway in lat. 60° N. . . .

"Another example of a rounded block, weighing above thirteen pounds, had been previously noticed in the 'chalk with flints,' . . . in a pit near Lewes. Attached to it was *Spondylus lineatus*, with serpulæ and some bryozoa. It had evidently been rolled before transportation, and before the serpulæ had fixed themselves on it." *

If the presence of the Mosasaurus, in lat. 51° N., be considered conclusive evidence that the climate prevailing there in the Chalk period was the same as is now seen twenty degrees of latitude further south, or about that of Alexandria, Egypt, whence these masses of floe-ice, evidently of Norwegian origin, with their freights of rock, sand, and gravel, — ice which under the assumed circumstances could not have been generated south of the latitude of Iceland? Icefields and icebergs drifting about in the tepid waters of a sub-tropical sea, or one of corresponding temperature, would indeed afford a most extraordinary spectacle, quite out of the usual course of nature. Again, if this and the other extinct saurians mentioned were reptiles of a purely tropical type, how can their presence be satisfactorily accounted for in latitudes exhibiting the phenomena of snow and ice?

* Prin. of Geol., vol. i. p. 216.

The records of the Cretaceous period, therefore, as far as yet deciphered, show the same *appearance* of abnormal warmth which we found characteristic of the several divisions of the Tertiary age of the world. The northerly range of the reptilian forms of the Chalk indicates, plainly enough, a lapse of time within that period, marked by a more equal distribution of the solar influence over all the earth than now prevails; while the presence of floating ice in the surrounding waters of the British Isles attests to an interval of sharply contrasted seasons, or to a Cretaceous Glacial period. With these proofs of climatal vicissitudes, the futility of attempting to assign a stated, invariable temperature to the whole Cretaceous period, is obviously manifest; and the legitimate conclusion to which they lead is, that, like the other geological periods, this, too, has experienced the same cycle of climatal changes, the result of the same cause, namely—a half revolution of the earth transverse to its diurnal rotation.

Climate of the Oolitic and Triassic Periods.—Lyell includes the consideration of the climatal condition of the world, during these two geological epochs, under a single head; and having attempted, as we have seen, to show that their successor, the Cretaceous age, was characterized by an undeviating uniformity of climate, so now he endeavors to make it appear that the same invariability of temperature — somewhat warmer, perhaps — prevailed during the five or six millions of years of the Oolitic and Triassic periods. He remarks that zoologists and botanists are very generally agreed as to the warmth of European latitudes during these times. "The vegetation of these periods," he says, "consists chiefly of cycads, conifers, and ferns. Professor Heer remarks, that the tree which is most common in the Upper Trias in Switzerland has a near affinity to a living African species of Zamia, and M. Adolphe Brongniart had long before expressed his opinion that the plants of the secondary periods favored the hypothesis of a climate like that of the West Indies. The same genera, and, to some extent, the same species of ammonites and some other shells proper to oolitic strata in Europe, occur also in formations of the same age in India, as, for example, in Scinde and in Cutch, lat. 22° N. In a northerly direction the same formations reach within 13½ degrees of the pole, as was shown by the fossil specimens brought home by Sir Leopold McClintock. Among these the Rev. Samuel Haughton recognizes a species closely allied to the *Ammonites concavus* of the Lower Oolite which was found at Prince Patrick's Island, lat. 77° 10′ N. In Cook's Inlet also, lat. 60° N., several ammonites of jurassic types, if not species, were obtained, and *Belemnites paxillosus*, a British liassic fossil. But what is far more remarkable, remains of a large ichthyosaurus of liassic type were brought from an island in lat. 77° 16′ by Sir Edward Belcher. They have been described and figured by Professor Owen, and as some of the vertebræ were 2½ inches in diameter, the animal must have been of considerable size. More recently, in 1866, the remains of ichthyosaurians were found by the naturalists of the Swedish expedition, in strata of jurassic age in Spitzbergen in the still more northerly latitude of 78° 30′." *

These remarkable and significant facts are explicable alone under our proposed new system. They point unequivocally to portions of time within each of the two periods under consideration, when the same analogy of climate between tropical and temperate, and temperate and polar latitudes prevailed, as we have already seen marked a portion of the Miocene epoch. We have now to account for the repeated occurrence of something analogous to a tropical climate at the pole, as well as for a generally prevailing uniformity of climate throughout all latitudes. And how are we to do this? No one will deny that the form of the earth during these two periods was, as it now is, spherical, intercepting in consequence the same amount of the sun's light and heat, or will assert that the inherent properties of solar rays were then different in any respect from what they now are. Those rays, meeting the earth's surface vertically, or perpendicular to the plane of the horizon, or any approximation thereto, would, then as now, produce a high temperature, or a tropical climate. Then as now, any part of the earth's surface subject during a part of the year to a total deprivation of the solar influence, and, during the remainder, receiving it only under a large inclination, must experience a frigid climate. Now, if the inclination of the terrestrial axis of diurnal rotation

* Prin. of Geol., vol. i. p. 217.

was the same in Oolitic and Triassic times as at present, the sun's rays would meet the surface of the earth at the same degrees of inclination in the different latitudes, and the same varieties of climate would prevail. Any analogy of climate, therefore, between equatorial and polar latitudes, would be as utterly impossible then as now; for any change in the intensity of the solar heat would affect all latitudes in the same relative proportion. The similarity of climate, therefore, which the facts indicate as having prevailed, from the equator to the poles, at times during the Oolitic and Triassic periods, is inconsistent with, and wholly impossible under, the present status, and can only have been the result of the cause we have assigned as producing like climatal conditions in subsequent geological epochs.

It is unnecessary to recapitulate here the arguments employed in former instances going to show the absurdity of attempting to account for a warm climate at the poles under the present status. There is no difficulty in estimating the precise effect in other parts of the world of such an augmentation in the heat of the sun as would be required to raise the annual mean of temperature at the poles forty or fifty degrees, or what would be needed there to develop and support an exuberant vegetation and a large abundance and variety of animal life. The geological record gives no indications whatever of the periodical searchings to which middle and low latitudes must have been at such times subjected, and the only rational method of accounting for these seasons of polar warmth, is in assuming them to have been the result of a change in the position of the earth with regard to the sun, such as would admit of those localities receiving, for a part of each year, the direct influence of the sun's rays. These hot polar summers would, it is true, be contrasted by winters of more or less severity, but cold winters are a comparatively slight check to the development of life, either animal or vegetable, when they alternate with seasons sufficiently hot and productive.

Lyell asserts that the abundance and variety of reptiles implies warm climate. The proposition is doubtless true so far as this: that the development of these animals depends on the prevalence of a tropical, or sub-tropical temperature, during either the whole or a *part* of the year. We have no right, however, to assume, even in the case of living reptiles, that, with their well-known power of burying themselves in the earth and hibernating in a torpid state during the season of cold, that they cannot endure winters of considerable severity, if such winters alternate with summers of excessive heat, and still less strictly is the proposition applicable to the extinct species of the secondary age, of whose habits we, in reality, know so little. There may have been those among all these numerous tribes, for all that we know to the contrary, capable of enduring continuous cold; and all of them, as they continued in existence, like the extinct mammalian orders of later times, throughout whole geological periods, must, like them, have been endowed with the capability of adapting themselves, through self-modification, to the ever-changing conditions of life consequent on continually varying climate.

In alluding to the diversified form and number of the reptiles of the Oolite, Lias, and still older Trias, he states that "the number of marine genera alone of this class exceeds fifty, while that of the fresh-water and terrestrial species, including those of aërial habits, is almost as great as that of the tribes which peopled the sea. Some of these were more highly organized than any animals of the same class now living, as the Belodon, for example, of the Upper Trias, a saurian about the size of the largest living crocodile, but which belonged to the extinct order of Dinosaurians. Hermann von Meyer ascertained, in 1865, that it possessed breathing apertures or spout-holes like the whale, so that we might imagine it to have been capable of sustaining a cold climate were it not associated with many reptiles of lower grade, as well as with shells, corals, and plants which bespeak a high temperature."* The eighty reptiles derived from the Trias of Germany, described by Hermann von Meyer, belonging entirely to extinct orders, but all of which, according to Owen, display affinities (structurally we must add), more or less decided to living families of the same class, the representatives of the still existing crocodilian and chelonian orders in the overlying liassic and oolitic groups, together with the

* Prin. of Geol., vol. i. p. 218.

four extinct orders the Pterosaurs, Ichthyosaurs, Plesiosaurs, and Dinosaurs, exhibit various grades of organization, and, it is asserted, "the analogy of the living creation is strongly in favor of their having flourished in a climate in which the heat was considerable during part of the year, and the winter brief and never severe."

Certain tribes of animals may display, in osteological structure, more or less decided affinities to other families of the class to which they may be referred by naturalists, and yet be related to such families only in the most distant manner, being constituted in all other respects, physiologically and otherwise, totally distinct from them. It is extremely improbable that the highly organized reptilian forms of the secondary age have degenerated into the inferior orders of the same class now inhabiting the earth. Such an assumption would be contrary to the obvious, ever-accelerating progress which every department of the organic world has displayed from the beginning, towards more and more highly organized states; but inconsistent as it would be with the general plan of nature, it cannot be avoided in the attempt to establish any close relationship between living reptiles and those of the Mesozoic age of the world under consideration. If, on the other hand, we look for the descendants of these last among the mammals of to-day, and view living reptile types as the offspring of lower and less progressive secondary forms, we can clearly see how slight is the degree of consanguinity or relationship that really subsists between living reptiles and the extinct Oolitic, Liassic, and Triassic orders above mentioned. This principle is not necessarily restricted to the animate portion of the organic world, but is equally applicable in the case of the vegetable kingdom. Mesozoic cycads, conifers and ferns may be considered, generally, as the progenitors of the superior vegetation of later times, while living types of the same classes are the descendants of less advanced secondary forms. If this hypothesis is tenable, what dependence is to be placed on inferences in regard to climate derived from an assumed relationship between ancient and modern forms, such relationship being predicated on a no more tangible basis than that afforded by a mere chance resemblance between the two in a single particular? The respiratory organs of the Belodon, the warm, shaggy covering of the mammoth, the transportation of masses of rock, and the striation and polish upon them, resulting from the buoyant power and grinding action of ice, are facts of far more value in determining the exact climatal condition of ancient geological epochs than any possible conclusions founded upon such variable and inexact data.

Were we to admit, however, that the affinity between ancient and modern reptile forms extends in the same measure to physiological constitution and habit as it does in osseous structure, even this hypothesis is not inconsistent with the idea that the former were able to endure winters of considerable severity, provided the summers were hot enough to meet the demands of their nature. At present, wherever upon the earth's surface this class of animals are scarce or are entirely wanting, the summers are always short and cool, whatever may be the character of the winters. Terra del Fuego, the woody region north of the Straits of Magellan (between latitudes 52° and 56° S.), the Falkland Islands, and the arctic regions are cited as instances of localities nearly or quite destitute of reptile life, and in all of them are the summers as above described, and to this fact we are entitled, doubtless, to ascribe the scarcity and absence of these animals.

But it is needless to pursue the theme further. It is indeed true that there was an interval of time in each of the three periods in question, when the climate of Middle Europe was the same as that described by Lyell as prevailing invariably throughout them all; that is to say, when the heat was considerable during part of the year, and the winter brief and never severe. There was another interval in each, when a universal uniformity of climate according to latitude prevailed all over the world, the result of perpetual equinox; the temperature of any part thereof being determined by its distance from the equator. Still another portion of each was characterized by the sharp contrast of seasons consequent on a large degree of obliquity of the terrestrial axis of diurnal rotation.

In proof of the occurrence of the latter conditions, or glacial intervals within these three epochs, we have the evidence afforded by the disposition of portions of the movable material of the earth's surface in regular series of layers or strata, such disposition being effected through the instrumentality of meteorological forces in a high state of activity, consequent upon great extremes of heat and cold; and, also, the direct testimony of erratic boulders occurring, as we shall hereafter see, in the conglomerate of one of these geological periods. All these climatal states and the gradations between them — these climatic cycles and the ever-changing conditions of life depending upon them — were experienced by the organisms of the times. These organisms, possessing, in common with those of all other ages of the world, a certain innate plasticity of constitution, by which they were enabled, through self-modification, to adapt themselves to such changes, were slowly but continually varying in habit, and more slowly still, perhaps, in structure. We may view each individual instance of variation as an effort of the animal or vegetable economy, to accommodate itself to some change in its surroundings. The variation having been accomplished and its immediate office performed, if, when the conditions inducing it were past, it should still, on the whole, prove beneficial, it would, according to the law of natural selection, be perpetuated; if not, it may be supposed to gradually give way to more useful modifications. Here, then, in brief, is the fundamental principle of the doctrine of evolution, the great law through the operation of which low primitive forms of life have been gradually improved into the superior types of modern times, and in this light we behold in the reptiles of the Secondary age an important link in the grand chain of progressive development of organic life.

Triassic Glacial Interval. — Lyell thus briefly alludes to the phenomena indicative of this geological event: "The great size of some fragments of rock in the New Red Sandstone, probably of Triassic age, in Devonshire, has led Mr. Godwin-Austen to refer their transport to ice-action; but this opinion has been controverted by Mr. Pengelly, who has shown that such masses may not have travelled far, and are such as might have been moved by breakers beating against a wasting cliff." *

* Prin. of Geol., vol. i. p. 221.

If we now again recur to Lyell's comparison cited on p. 83, and also to his remarks on p. 97, in which the obvious tendency of the superficial changes in progress during a glacial period to obliterate and destroy the work of its predecessor is vividly illustrated and strongly enforced, no surprise need be experienced if, in the present stage of geological investigation, but one of these three epochs has been found to exhibit traces of glacial action, and these traces even not of the most decisive character.

It should also be borne in mind that every instance in which the phenomena of glaciation occur in the early geological periods is in direct contravention to his hypothesis of a universally prevailing uniformity of climate, and of a (retrogressive), gradually increasing terrestrial heat. Thus, in a case where there was no alternative, we were informed that the theory of ice-action "has been resorted to most unwillingly;" and it may readily be supposed that in doubtful cases any expedient, however improbable, would be employed in order to avoid the necessity of having recourse to such theory.

But whether these Triassic boulders are true glacial erratics or not, is a question an affirmative answer to which is not essential in the establishment of the truth of our theory.

Supported as we are by so much other entirely unimpeachable testimony, proofs so various in kind, and drawn from so many different sources, all tending directly to the same end, we hold that the view which regards the rock-masses referred to, as ice-borne erratics, is much more justifiable than that assigning their transportation to the action of the waves of the sea.

Permian Climate.—Between the Triassic and Permian rocks there is a break, says our authority, which doubtless implies a great lapse of time, of which the records are wanting in that part of the globe as yet best known to the geologist. This missing link in the geologic chain, corresponding doubtless to a semi-transverse revolution of the earth, or a geological period, "constitutes the line of division between the primary and secondary, or between the Palæozoic and Mesozoic formations. The Permian rocks have been traced as far north as Petschora-land in Russia between lat. 65° and 70° N. They occur largely in Germany and England; and in North America have been traced as far south as Kansas and Nebraska, lat. 44° N."*

The fact that, among Permian fossils, a few forms more or less closely allied in structure to living types inhabiting warm countries are to be found in these northern locations, is held to "indicate the prevalence of a warm and moist climate throughout a great part of the northern hemisphere;" and these conditions are supposed to continue uninterruptedly during the whole epoch.

The genera Nautilus and Orthoceras, among Permian shells, and large reptiles which sometimes accompany them, belonging to a family called Thecodonts, which combine in their structure many characters of living crocodiles and lacertians, and, in general, the whole flora of the Permian formation, are the fossils adduced in support of the above conclusion.

If, in the present instance, the evidence brought forward is sufficient to justify this conclusion, and no reasonable doubt can be entertained but that tropical and sub-tropical climates prevailed throughout the whole Permian epoch in the northern latitudes above mentioned, when we come to encounter, in the same formation, the most legible and authentic traces of a glacial interval, as described in the quotation following, there seems to be no resource remaining but to abandon as futile all attempts to investigate the climatal status of this geological period.

Permian Glacial Interval. — Lyell records the "supposed" signs of ice-action in the Permian period as follows: "Professor Ramsay, in an able memoir published in 1855, gave an account of observations made by him on a brecciated conglomerate of Permian age, in Shropshire, Worcestershire, and other parts of England, which had led him to infer the action of floating ice in the seas of that remote period. His arguments are founded on the following facts: — the fragments of various rocks imbedded in these breccias are often angular, and of large size, some of them weighing more than half a ton; they are very often flat-sided, and have one or more of their surfaces polished

* Prin. of Geol., vol. i. p. 222.

and striated. They are generally enveloped in a red unstratified marl, in which they lie confusedly, like stones in boulder-drift. In some cases it can be demonstrated that the nearest points from which these stones could possibly have been conveyed are the mountains of Wales, more than twenty, thirty, or even fifty miles distant; and it is inferred that the only way in which they could have retained their angular shape, after being transported so far from their original position, is by being carried by floating ice. Some of the specimens also taken by the Professor out of the breccia, and now exhibited in London, in the Jermyn Street Museum, have their surfaces rubbed, flattened, and furrowed, like stones subjected to glacial action. One of the most characteristic of these specimens was obtained from a spot about six miles south-east of Bridgenorth, near the village of Enville, in Worcestershire. The fragment is six inches in its longest diameter, consists of hard, dark Cambrian grit, with a smoothed surface, exhibiting parallel sets of striæ in more than one direction, a newer set crossing the older one. I am fully satisfied that such fragments have been taken out of the breccia, and the explanation offered by Professor Ramsay appears to me the most natural, indeed the only one in the present state of science which can be suggested." * Among other observations, after remarking upon the great dearth of fossil remains in the Permian conglomerates of Central England, a universal characteristic of glacial deposits, he adds that "Professor Suess, who has studied the Permian conglomerate or Rothliegende in various parts of the Alps, says that it shows signs of great denudation of pre-existing land by the large quantity of quartz pebbles which it contains."

The hypothesis of a Permian glacial interval is thus fully substantiated, not only by unmistakable traces of floating ice, but also by all the other manifest effects of glacial action, such as the polishing, channelling, and transportation of rocks, and the denudation of extensive land surfaces accomplished during the time of greatest contrast between summer and winter.

Our only refuge, therefore, from the inconsistency of supposing that all the phenomena incident to a geological

* Prin. of Geol., vol. i. p. 222.

glacial period could be in progress during a time of universal warmth, such warmth, under the assumed prevalence of the present terrestrial status, and inclination of the sun's rays, being to a great extent equalized in all the latitudes of the earth from the equator to the poles, — a manifest physical impossibility, — is that afforded by the theory of the transverse rotation of the earth. Availing ourselves of this refuge, we can accept every item of inorganic evidence at its full value, there being not a particle of the same requiring to be explained away or modified in any respect from its obvious legitimate bearing, to be in perfect harmony with such theory; and, indeed, after due allowance for a certain degree of variability, not necessarily of large amount, the same may be said of that derived from the organic world, for both classes of testimony, when viewed from the proper stand-point, unequivocally attest the permanent continuity or rotatory nature of the well-known motion of the earth assigned as the primary cause of the climatal mutations under consideration.

Carboniferous Period. — In relation to the climate of the Carboniferous period, we are informed "That botanists have considerably modified the ideas which they originally entertained respecting the tropical temperature supposed to be indicated by the fossil plants of that era. The fruit called Trigonocarpon, occurring in such profusion in the coal measures, was at first referred to the palm tribe, till the discovery of more perfect specimens enabled Dr. Hooker to decide that it was not a palm, but more probably belonged to a taxoid conifer, somewhat like the Chinese Salisburia." * This leads us to remark upon the inherent, essential difference between the climates of polar and of equatorial latitudes, and the effect of this difference upon the organisms peculiar to each. Among all the climatic vicissitudes of the former, incident to a half transverse revolution, there is no approach to that now experienced within the tropics; for, although there is an interval characterized by hot summers, no year in the whole cycle is exempt from a winter as rigorous as must follow the entire absence of the sun. Polar climate, in general, is, then, a frigid climate, broken by occasional intervals of warm

* Prin. of Geol., vol. i. p. 224.

summers. On the other hand, in the whole cycle of climatal variation experienced at the equator, at no time is there any approach to the conditions of an arctic climate, and equatorial climate, in general, is a torrid one, occasionally interrupted by comparatively brief spells of semiannual cool seasons.

Primary climatal variation does not proceed in an irregular, indeterminate manner, exhibiting indiscriminately every phase of change intermediate between complete reversal of the extremes, sometimes producing a torrid climate within the polar circles, and at others an arctic climate at the equator, with every possible combination of both in all parts of the earth — conditions such as might possibly be consequent on the operation of chance combinations of dissimilar agencies; but such variation, as may be learned from our retrospect, takes the form of regularly recurring uniform cycles, these cycles exhibiting an essential and fundamental difference in equatorial and polar countries, owing to the different circumstances under which the producing cause acts in the two localities. It is far from unreasonable to suppose that this inherent difference in the whole range of climatal change peculiar to the tropics, and that which has prevailed within the polar circles, has been the grand agency which has evolved out of a class of low primordial forms, originally common to the whole world, the peculiar fauna and flora of each of those divisions of the earth's surface. It may be assumed that continued repetitions of the polar climatal cycle have originated the arctic fauna and flora within the polar circles, at the same time that a similar succession of equatorial cycles developed, from the same progenitors, the classes of organic forms peculiar to the tropics. Thus the whole assemblage of organisms, of any part of the terrestrial surface, at any epoch in the world's history, are not to be viewed as the product merely of the conditions prevailing at that time and place, but are to be regarded generally, as the exponent of the whole sum of progress the living world had then and there made towards a more highly organized state.

The great numerical preponderance of ferns over all other forms of vegetation, in the Carboniferous era, is not, in this light, indicative merely of any particular climatic state, nor of the absence of more highly organized, and therefore more successful, competitors in the struggle for life. The Carboniferous flora did not spring up, because these and perhaps other favorable conditions happened to prevail at the time, but it was the result indicating the progress which vegetation had then made from the inferior types of earlier periods ; just as the reptiles of the Secondary age represented, in general, the total advance the animate world had at that time made from its inferior primordial progenitors.

It must be conceded, however, that the exuberant growth and the accelerated progress in development characterizing the vegetation of the Carboniferous period, imply highly favorable conditions. Whatever those conditions may have been, it is certain that they prevailed in all of the temperate latitudes, except those near the tropics, and, most probably, over the whole extent of that part of the earth's surface included within the polar circles. For says Lyell, —

" As to the geographical range in the northern hemisphere of this ancient flora, it is already ascertained that it extends from Alabama in the United States in lat. 30° to the arctic regions, while it has been traced in Europe from Central Spain in lat. 38° to Scotland in lat. 56°. In the arctic regions it was first observed in Melville Island, in lat. 75°, during Capt. Parry's expedition. The plants then collected were examined by the late Dr. Lindley, who recognized them as true fossils of the ancient coal. The original collection has unfortunately been lost, but among other fossils since brought from the same island by Sir Leopold McClintock, Heer has recognized ferns of the genus *Schizopteris*, a form characteristic of the ancient coal. Middendorf found *Calamites cannæformis* in a very high latitude near the mouths of the Lena. Von Buch has described strata of the Coal period containing characteristic marine fossils in Bear Island, lat. 74° 36′ N., midway between Spitzbergen and the North Cape, in about the same parallel as Melville Island ; and from associated rocks of the same age and in the same locality Heer has received as many as fifteen species of plants well known as occurring in different stages of the European Carboniferous formation.

"After what was said at p. 201" (cited on page 92 of this work), "of the spread of the Miocene flora over the arctic regions, and its near approach to the North Pole, the reader will feel no surprise at finding in times long antecedent there was an equally vigorous vegetation in the same latitudes. Moreover, the coal plants were of different genera, and some few of them perhaps of different orders, from any now existing, and they may therefore have been endowed with a constitution enabling them to accommodate themselves to a long polar night." *

Having, in the concluding sentence of the foregoing quotation, admitted, in substance, the truth of our doctrine concerning the nature of organic evidence, he proceeds to remark that even tropical plants will flourish wherever a requisite degree of warmth is maintained, even though deprived in a large measure of the bright light of the sun; and as an instance in point, allusion is made to the luxuriant growth of tropical plants in the hot-houses of St. Petersburg, in lat. 60° N., where they change the perpetual equinox of their native regions for days and nights which are alternately protracted to nineteen hours and shortened to five. The obvious intention is to show that the Carboniferous vegetation flourished under the present terrestrial status, and inclination of the solar rays, sufficient light being received directly from the sun to satisfy the requirements of vegetable growth, while the requisite heat, which, of course, must have originated from the same primary source, was conveyed to those high northern latitudes in some indirect, roundabout way.

The Carbonaceous flora constitutes a distinct assemblage of plants, and its presence, in any locality, evidently implies the prevalence there of certain conditions of growth depending chiefly on climate. The geographical range of this flora, extending in the northern hemisphere, on both the eastern and western continents, from the neighborhood of the tropic, nearly, if not quite, to the pole, indicates with remarkable precision the prevalence of the same climatal analogy, during a part of the Carboniferous era, which, we have before seen, marked a portion of the Miocene and other geological periods. As urged in those former instances, such analogy between the climates of all

* Prin. of Geol., vol. i. p. 225.

these different latitudes is wholly at variance with, and impossible under, the present relative positions of the earth and sun. As before argued, this analogy cannot possibly have been consequent on solar variability; for any change whatever in the potency of the solar influence must, in the nature of things, be felt in equal relative proportion over the whole terrestrial surface, and it is equally clear that there is but one conceivable method in which it can have been brought about. The equal, or nearly equal, distribution of the whole annual amount of the sun's light and heat over the entire surface of the earth (for it is impossible that the northern hemisphere could enjoy such distribution, and not the southern), which the climatal analogy that we know prevailed in both the Miocene and Carboniferous periods implies, can only have been produced by such changes in the mode of presentation of the terrestrial surface to the sun as are involved in the transverse rotation of the earth. These facts alone, therefore, vindicate the soundness of our position, and demonstrate the truth of our system beyond cavil.

The warm, humid, and equable climate which Lyell assigns to the Carboniferous, in common with the older geological periods, is precisely the one under which the accumulation of the vast beds of effete vegetable matter, subsequently transformed to mineral coal, would have been impossible. It is an undisputed fact that in warm climates, at the present time, the annual crop of dead vegetation rapidly and completely decomposes, and passes into the atmosphere in a gaseous form, the only trace of thousands of years of luxuriant growth being merely a darker shade of color imparted to a few inches of the surface soil. An exuberant vegetation might, therefore, flourish in a warm and equable climate, not only for one or six geological periods, but any other conceivable amount of time, and yet leave not a trace of mineral coal to succeeding ages. On the other hand, the conditions resulting from sharp contrasts of seasons may readily be supposed to favor the deposition of vegetable matter in a state, and in the quantities, requisite to produce the abundant supplies of coal now existing in the bowels of the earth.

In view of the obvious rapidity of vegetable development during the era of the Coal, it is difficult to resist the conviction that favorable conditions, other than those strictly of a climatic nature, united with the latter in producing so exuberant a growth. It has been suggested that an excess of carbonic acid over the quantity now contained in the atmosphere was one of these conditions; and it is not impossible that other of the proximate constituents of plants, existing in the air in a gaseous form, were more abundant at that time than now.

In regard to the supposed excess of atmospheric carbonic acid during the Carboniferous period, Lyell endeavors to show that the quantity contained in the air is invariable, and that the vast amount annually absorbed in the various processes of organic development, is replaced by "gaseous emanations from the interior of the earth, which are most copiously given out in volcanic regions, and especially by volcanoes during eruptions. Carburetted hydrogen," he observes, "also escapes from beds of coal and lignite, and other fossiliferous strata, in which organic matter is decomposing; the same gas, evidently rising from great depths, is also evolved from rents in the granitic and other crystalline rocks in which there are no organic remains." *

The argument does not appear conclusive. Allowing that such process of compensation has been in operation ever since the Carboniferous era, it does not follow that it was in existence previous to that time. The estimate that ten times more carbon is locked up in a solid form in the ancient coal measures than all that is now contained in the atmosphere, is admitted to be far below the mark; and if so, it seems absolutely impossible that such an immense amount of carbonic acid could have been abstracted from the atmosphere during a single geological period, without affecting, in some degree, the proportional quantity of it relative to the other constituents of the same. If the Carboniferous period was the first to witness a general development of vegetation, the scarcity or absence of plant life in antecedent ages denotes that an efficient agency in removing from the air the subterraneous emanations, as well as a supposed original excess of this

* Prin. of Geol., vol. i. p. 226.

compound of carbon and oxygen, was wanted in those ages; so that, independent of the primal excess, the proportion may be supposed to have been continually augmenting. These considerations, however, appear to be of a superficial character. Evidently we must delve deeper to arrive at the root of the matter.

The quantity of carbon permanently locked up in the coal, vast as that quantity is, represents but a small fraction of the whole amount contained in the crust of the earth. Combined with oxygen, it enters largely into the composition of all kinds of calcareous rock, such as the chalk, limestones, corals, and other earthy carbonates which make up so large a proportion of the solid substance of the globe. Every novice in the science of chemistry is aware of the moderate degree of heat required to separate carbonic acid gas from all of these compounds. It follows, therefore, that if the earth has ever existed in an incandescent state, the total amount of carbon appertaining to it must have assumed, by union with oxygen, the gaseous form, and thus have constituted a part of the intensely heated and attenuated aerial envelope of the fiery liquid nucleus, which nucleus, it would seem, must have consisted of a homogeneous mass of basic substances.

The doctrine of the original fluidity of the earth involves, therefore, not only a primal excess over the present proportion of carbonic acid contained in the atmosphere, but also the fact that such excess included the whole of that compound now permanently fixed in the earth. It seems probable that the time needed to complete the cooling-off and consolidation of an exterior crust was short, compared to that required for the removal of such large excess from the air. Geologists are rapidly approaching the conclusion that *all* the calcareous rocks are of organic origin. Whether this is so, or whether certain varieties may have been formed by the direct chemical union of carbonic acid with the calciferous base, it is quite certain that the deposition of these rocks, and the burying up of organic remains, must have preceded the evolution of carbonic acid from the volcanic and other fissures in the earths crust; the emission of the gas being in consequence of its disengagement, by the internal heat, from the earthy carbonates and organic remains. In view, therefore, of the slow and complex process or processes by which, previous to the advent of vegetation, carbon was abstracted from the atmosphere, and fixed in permanent form in the earth, it seems perfectly safe to conclude that the present equilibrium between the amount of this element injected into the atmosphere, and that absorbed from it by vegetation, did not prevail until after the Carboniferous era.

Another cause which, in conjunction with the above, may have aided during the Carboniferous period, in the formation of the immense deposits of effete vegetable matter, is to be found, perhaps, in the character of the plants composing the flora of that time. Many of the ferns, it is said, attained arborescent proportions; and their lax tissue suggests great rapidity of growth. In the absence of any definite knowledge, generic or specifical of many of these plants, it is impossible to affirm, notwithstanding their size, that they were not annuals, attaining full development in a single season. It is quite probable that some of them, at least, were such; and with plants of this nature, a superabundance of plant food and favorable climate, it is not difficult to conceive that the annual product of carbon on a given area might equal that it now requires twenty-five, fifty, or even a hundred years to produce.

But whatever value may attach to the above considerations in accounting for the immense vegetable deposits of the Coal era, the rapid and luxuriant growth of the Carboniferous flora is a fixed, undoubted fact; and this enormous annual growth implies the potency of the solar influence. The coal measures have been aptly compared to reservoirs in which the surplus light and heat of the sun of a former time have been stored for the benefit of succeeding ages; and nothing can exceed in absurdity the idea that such vast supplies of the solar energy could accumulate in the absence of the sun. A torrid clime must have prevailed while the plants were growing, else there could have been no sueh luxuriant growth. A torrid climate, and that which could alone produce it, — a vertical or nearly vertical sun, — could not have ruled throughout the year; for in that case the vegetable matter would decompose as fast as it should lose its vitality. The tropical summers must, therefore, have alternated with winters of more or less severity. Wherever coal is found, whether in low temperate latitudes or near the pole, during the time of its depositions in the form of dead vegetable matter, the year must have been divided as above stated. There is positively no rational mode of avoiding this conclusion; and, admitting its validity, we have the means of accounting, in a natural manner, for the preservation of the annual crops of dead vegetation in a condition favorable to its subsequent transformation.

Let us suppose that the river-valleys of the antecedent or Devonian system, on the approach of the Carboniferous Glacial interval, began to fill up with alternate layers of ice and earthy material, or ice and vegetable matter, in the manner heretofore described. The sun of each succeeding spring would liquefy the ice above the uppermost layer of earth or vegetable matter, and the river-beds may be supposed to assume in summer the appearance of extensive fens or marshes. The plants of the coal, springing up from spores, seeds, and perennial roots, and deriving by far the larger portion of their sustenance from the atmosphere, would in a short space of time effectually screen the soggy surface from the rays of the sun, and prevent the warmth from penetrating into the ground to any considerable depth. In their rapid development they would soon assume an appearance justifying the description naturalists have given us of the vegetation of the period. Some of the conditions which, according to our theory, must have prevailed in the Coal era, judging from present analogies, seem to be unfavorable to rapid growth. But we are to bear in mind the fact that the plants of that time were as closely adapted to the conditions of life by which they were surrounded as are those of the present, and also, that however closely a plant of the Carboniferous flora may resemble, in structure, some living type that can flourish only under diametrically opposite circumstances, the fact cannot militate against our theory; for in our ignorance of the habits of the carboniferous plant, we do not know but the very condition that must inevitably prove fatal to the modern type was the very one on which depended the existence and rapid growth of the other.

Having arrived at maturity, the plants of our carboniferous swamp are, on the approach of winter, cut off by the frosts, to make room for the succeeding crop. The dead vegetation, falling down, becomes saturated with moisture, and covered with quick-growing mosses. A temperature sufficiently low to insure its preservation is maintained through the hot season by means of the underground ice deposits, and the dense shade afforded by the foliage of the new growth. This process repeated annually for thousands of centuries, interrupted occasionally by chance inundations, consequent, perhaps, on the exceptional exposure of some vast body of ice to the rays of the summer sun, the effect of which would be the interposition of a stratum of earthy material in the vegetable deposit, seems most in accord with the phenomena of the Carboniferous period; and the same phenomena, indicating unmistakably an analogy of climate between low temperate and polar latitudes, furnish us with an unanswerable argument, among many others, in favor of the theory of the transverse rotation of the earth.

Davonian Period. — Placing implicit reliance on evidence derived from greater or less degrees of structural resemblance between the fossils of the Triassic, Permian, Carboniferous, Davonian, and the still older Silurian and Cambrian systems, and still living organisms, — a species of evidence we hold to have become, at this stage of our retrospect, entirely worthless as a means of determining the climatal conditions of those remote eras, — Lyell arrives at the conclusion that "a similarity of conditions in regard to temperature prevailed throughout the whole of these six periods." * These supposititious conditions, as we have noted in the cases of the first-named three, are assumed to result in a climate described as "warm, moist, and equable," and the same conditions are assigned to the Davonian, Silurian, and Cambrian epochs.

It seems proper, in this connection, to reassert the fact that such uniformity of climate which, from the terms employed, we are to understand as prevailing generally throughout the entire year, and over the whole earth, is, and in the nature of things must be, a manifest physical impossibility. Under the present relative position of the earth with regard to the sun, the annual amount of solar

* Prin. of Geol., vol. i. p. 232.

light and heat transmitted to the earth must inevitably be distributed over its surface in the same manner, exhibiting everywhere the same variety of seasons, such as we now behold. At no point on the earth's surface can there be the least variation in the relative length of the days and nights, as now observed during the several seasons of the year, and the rays of the sun must be received at the same degrees of inclination in those seasons that they now are. Any change from the present climatic condition, any approximation towards a general uniformity of climate between the different latitudes of the earth, is, therefore, as we have stated, a physical impossibility under the present status. So also a change in the relative position of the earth, such as would bring the equator coincident with the ecliptic, while it would entail perfect uniformity of seasons in all latitudes, and an invariable temperature to each, would at the same time exhibit the greatest possible contrast between the climates of equatorial and of polar regions. The nearest approach to a general uniformity of climate the earth can ever experience is when the equator becomes perpendicular to the orbital plane; but even then, although the annual sum of solar heat is equally distributed over its surface, the difference between the summer and winter temperature of each and every latitude must be very great; so that nothing analogous to what may properly be termed a universally uniform or equable climate can have prevailed at any time during the existence of the world. Our author's conclusion, therefore, concerning the climatic state of the earth during these six periods, appears untenable from any point of view we may choose to assume.

Glacial Interval of the Davonian Period.—Additional to the foregoing considerations, this hypothetical uniformity of climate assumed to prevail during nearly one half of the entire existence of the earth of which we have knowledge, is disproved by the direct, authentic evidence already adduced showing the intervention, both in Triassic and Permian times, of intervals of glaciation. That a similar interval occurred in the era of the Old Red Sandstone is evident from the following citation, in which are described the earliest signs of ice-action as yet discovered.

"The Rev. J. G. Cumming, in 1848, in his History of the Isle of Man, compared the conglomerate of the Old Red Sandstone to 'a consolidated ancient boulder clay;' and more recently (1866), Professor Ramsay has pointed out that the conglomerate of the same age seen at Kirkby-Lonsdale, and Sedbergh, in Westmoreland and Yorkshire, contains stones and blocks distinctly scratched, and with longitudinal and cross striations, like the markings produced by glacial action. I have myself examined this rock, and have seen blocks taken from it which exhibit such markings, some of them undistinguishable from those which I have observed on blocks taken from beneath a glacier."*

These facts, when allowed their proper scope, directly controvert the theory of a general uniformity of climate; and an attempt is therefore made to show that the markings may have been produced by various movements to which the conglomerate alluded to has been subjected, and the great pressure consequent on its being buried under thousands of feet of carboniferous strata.

The extreme improbability that the causes assigned could produce identically the same effects as those resulting from glacial action is evident at a glance; and even the author himself, if we may judge from the concluding sentence of the paragraph, seems aware of the weakness of his attempted evasion, and more inclined to believe in the glacial origin of the markings than in the other hypothesis. The sentence alluded to, a single word of which I have put in Italics, is as follows: "More evidence, I think, must be obtained before we can feel *perfectly* convinced that the markings in question have had a glacial origin."

If geological investigation in this particular direction shall become more general, and not, as heretofore, be confined to the desultory efforts of a few individuals, important additions will, no doubt, be made to the present amount of evidence of this nature; but from the apprehensions expressed by Lyell in the fore part of our retrospect of its probable scarcity, and the reasons given for indulging those apprehensions, it would almost seem as though the

* Prin. of Geol., vol. i. p. 229.

evidence above recited is all we can reasonably expect to obtain relative to a Devonian Glacial period. The general character of the Isle of Man conglomerate, and the glacial striations on the imbedded stones in those of Westmoreland and Yorkshire, attest with a considerable degree of precision the reality of such a geological event. The same facts taken in connection with other corroborative testimony authorize us to accept it as an absolute certainty.

When we consider the fact that the conglomerates of all ages are composed of materials identically similar to those of the more modern glacial and alluvial deposits, and as every practical geologist must have observed, that these materials are disposed throughout such rocks in precisely the same manner they are in the so-called post tertiary drift, in many instances showing the same varieties of stratification, — regular, irregular, bent, broken, and contorted, — and in others exhibiting a total absence of any signs of stratification, it seems impossible to resist the conviction that such conglomerates are nothing more or less than the concreted drift formations deposited during the glacial intervals of anterior geological periods. This suggestion, which appears to be perfectly natural and consistent, seems to merit the attention of geologists.

Allowing for an intervening geological period between the Trias and the Old Red Sandstone, the entire records of which we may suppose to have been lost, more than twenty-five millions of years ago glacial processes similar to those which characterized the latest Glacial period within the British Isles were in progress in the same localities. The presence of glaciers implies a determinate degree of cold. The climate, therefore, of those Isles must have been, at least, approximately the same during the glacial intervals of both the Pliocene and Devonian epochs, or cycles; and if they were the product of the same cause, and the effect of such cause at any given point in each was the same, it follows that the cause was acting in both instances under similar conditions, and we may reasonably infer that the two cycles were alike in all their parts. The conclusion, then, is, that there has been no material variation in the mean temperature of the globe for the last twenty-five million years, at least, and that, during the

Devonian epoch, when the earth's axis of diurnal rotation assumed its present degree of inclination, the same annual mean of temperature, and the same climatic vicissitudes incident to the several seasons, prevailed in all the latitudes of the earth, such as those which characterize the present time. Do we not here obtain a glimpse of the true stability of nature, as indicated in the permanence of the present order of arrangement of the solar system, and in the invariability of those fundamental conditions under which alone the commencement and continuance of life upon our planet has been possible?

In concluding this branch of our subject, it may be remarked that the facts elicited in the course of our inquiry concerning the climate of the world in former ages, show that all the innumerable changes to which it has been subjected, far from being chance mutations depending on the agency of irregular and capricious local and secondary causes, acting either singly or in conjunction, proceed, on the contrary, in regular, determinate similar cycles, necessarily implying the operation of a grand controlling agency, uniform and constant in its nature, the effect of which has been always relatively the same throughout the whole earth, from the most remote periods of antiquity of which we have any definite knowledge, down to the present time. The reality of these climatal cycles, and their connection with and dependence upon the transverse rotation of the earth, is shown in the case of the present, or Pliocene cycle, by a decrease of mean temperature in polar regions; such decrease, under the conditions there obtaining, proceeding at a rate so rapid as to be perceptible in its effects upon the welfare of the inhabitants of those regions, almost from century to century. In somewhat lower latitudes the same decrease of temperature, necessarily requiring a much longer period of time to bring it about, is indicated by the migratory movement southward of the molluscous fauna of the Norwegian seacoasts. In Middle Europe the same decrease of temperature, although in itself less absolute, and depending more on the occurrence of colder and colder winters than on concentration of the solar heat towards the equator, is apparent in the extended southerly range of the reindeer and other northern animals during the intermediate and earlier portions of the age of Stone—a change requiring a still longer interval of time. Thenceforward the cold appears to increase in a ratio corresponding to the increase in the inclination of the earth's axis of rotation, until we encounter the phenomena of the Inter-Glacial period, at which time a most remarkable uniformity of climate, or "climatal analogy," is seen to have prevailed from the northern tropic to the immediate neighborhood of the north pole, or, as far as known, throughout all latitudes. Again the intense cold appears to have prevailed, gradually decreasing in severity during the earlier portions of the epoch in the same uniform manner that characterized its increase during the later divisions of the same. The apparent rise in the terrestrial temperature seems to continue until we encounter the Upper Miocene, at which time, as far as it is possible to judge, the annual round of seasons incident to the several latitudes had given place in each to an invariable temperature, or a perpetual equinox, conditions that can result only from coincidence of the earth's equator with the orbital plane.

In the next, or Miocene cycle, it becomes impossible to follow the graduated changes from one marked climatic state to others, so perceptible in the Pliocene, or present era; but the evidence pointing to the prevalence of those diverse states, such as must invariably follow the several positions assumed by the earth with regard to the sun in the course of a half transverse revolution, is conclusive. Indian fossils of Upper Miocene age testify to a concentration, in that era, of the solar energy in the direction of the equator, and a general uniformity of climate. The intervention in Middle Europe of a Miocene Glacial period is a fact beyond question, while the inter-glacial interval of the same cycle is unmistakably indicated by the known climatal analogy which extended from the West Indian Islands, at least, to the immediate vicinity of the pole, and also by the extreme northerly range of the Miocene flora. The records of the next antecedent or Eocene period establish with corresponding certainty that a glacial interval constituted a portion of that cycle, and the character of the Eocene fossil fauna and flora of Middle Europe attests no less conclusively the prevalence of an Eocene Inter-Glacial period, while the proofs of an Eocene era of repose, although less pronounced, are not wanting. One point in the Cretaceous revolution bears the obvious marks of still another Glacial period, and the fossils of another portion of the same epoch indicate another repetition of climatal conditions resulting from equal distribution of the whole annual amount of the solar influence throughout all latitudes. In the Oolitic, Liassic and Triassic divisions of the Secondary age, and in the Permian, Carboniferous, and Devonian epochs of the Palæozoic, the same round of phenomena is in every instance repeated with more or less distinctness, each geological period being marked by one or more radical, highly-contrasted deviations from the climatal status which now prevails over the face of the earth.

Obviously, the position the earth may assume with regard to the sun, in its annual revolutions about the same, by regulating the proportion of the annual supply of the solar influence each latitude shall receive, and also the manner in which such proportion shall be distributed over the year, determines, in general, the terrestrial climate. When, even in a single instance, therefore, the geological record shows the prevalence of a mean annual heat in the vicinity of the pole sufficient to develop an abundant and diversified animal and vegetable life, or when, by the same means, we arrive at a knowledge of the fact that at a certain period in the terrestrial history an approximately identical climate prevailed from the tropic to the pole, either instance involves a radical change in the manner under which the terrestrial surface was presented to the sun from that now obtaining. If the record exhibited but a solitary instance of each of these and other widely differing climatal states, while such deviations could only be rationally attributed to changes of position of the earth's axis, such changes might readily enough, perhaps, be supposed the results of chance agencies; exceptional departures from the ordinary course of nature; but when we behold them occurring almost invariably in each and every geological period, there seems to be no escape from the conclusion that, whatever their producing cause, it must have been constant in nature, operating continually throughout all the ages of the world. Seven of the thirteen geological revolutions have left legible and most conclusive traces of their intervals of glaciation engraved upon the enduring rocks; and the same proportion witnessed, in polar countries, an abundant and varied fauna and flora, thereby attesting to the action of a vertical sun. The annual amount of heat received by the earth from the sun being invariable, or nearly so, this polar warmth necessarily implies a corresponding diminution of mean temperature in the direction of the equator, the same tending to produce an analogy or similarity of climate from the equator to the pole. An interval marked by this climatal analogy is distinctly traceable in no less than six of the geological periods. Repeated instances also occur in which a general uniformity of climate according to latitude has prevailed, with concentration of heat towards the equator. Not one of all these numerous deviations from the present climatal state of the world could possibly have taken place with the earth's axis in or approximating its present position. Viewing this array of facts, and the obvious bearing of such facts in the premises, it does indeed appear as though we might with perfect confidence rest the case upon the evidence furnished by the science of geology alone. So plainly perceptible are the effects of the transverse rotation of the earth upon climate, and through climate upon both the organic and inorganic kingdoms, that had no such motion ever come under the notice of astronomers, the facts presented would amply justify us in assuming its existence. We need not, however, rely in this case upon a mere assumption, whatever its degree of probability, for the observations of astronomers, extending over thousands of years, demonstrate the fact that the earth is subject to a movement which, if continued indefinitely, must determine in stated times complete revolutions of the same. But high astronomical authority refuses assent to the doctrine of continuity, and asserts the motion to be reciprocal or oscillatory. When this decision is seen to constitute the only obstacle to a perfect solution of the great problems of geogony, it certainly must be regarded with distrust; and all who are interested in the progress of human knowledge cannot but unite in insisting upon a thorough and searching revision of the methods through which such decision has been reached.

CHAPTER IV.

THE PLIOCENE GLACIAL PERIOD WITHIN THE TROPICS. — AGASSIZ' JOURNEY IN BRAZIL.

IDENTITY OF GLACIAL PHENOMENA OVER THE WHOLE EARTH, ACCORDING TO LATITUDE, IMPLIED IN THE NATURE OF ITS CAUSE. — PROPOSED OBJECTS OF PROFESSOR AGASSIZ' EXPEDITION TO BRAZIL. — DRIFT DEPOSITS OF RIO DE JANEIRO. — LETTER FROM PROFESSOR AGASSIZ TO PROFESSOR PIERCE DESCRIBING THE SAME. — SAME FORMATION OBSERVED AT VARIOUS POINTS ON THE ATLANTIC COAST, RIVER AMAZON, AND GENERALLY THROUGHOUT THE AMAZONIAN VALLEY. — ERRATIC BOULDERS OF ERERE. — GLACIERS OF THE SERRAS OF MONGUBA AND ARATANHA. — GLACIAL PHENOMENA OF THE ORGAN MOUNTAINS. — OBSERVATIONS ON THE FOREGOING FACTS.

THE facts thus far adduced as showing the climatal vicissitudes of former times have been derived chiefly from observations in the middle and higher latitudes of the northern hemisphere. This is the case, not because such evidences are wanting in other parts of the globe, but for the reason that the latter localities have not been subjected to such repeated and critical geological surveys as the former. It must indeed be confessed that our theory implies the presence of analogous phenomena over the whole extent of the solid portion of the earth's surface; that is to say, the same geological formation of any given latitude must present substantially the evidence of the same climatic changes, whether situated in the northern or southern, eastern or western hemispheres. If, for instance, under our theory, there has ever been a period or periods at the north pole when the semi-annual days and nights, representing respectively summer and winter, were determined, in the case of the day by the ascent of the sun from the horizon to the zenith, and in that of the night by its descent from the same to the nadir, like phenomena must have been contemporaneous at the south pole; and if, as another consequence of the same position of the terrestrial axis, the result was to produce an Inter Glacial period under the 45th parallel of latitude in the northern hemisphere, the southern must have experienced, at the same time, a similar effect under similar circumstances. The extreme relative disproportion of land and sea existing between the northern and southern hemispheres, must effectually preclude the idea of discovering in the southern half of the globe phenomena precisely analogous to those of Northern Asia, Middle Europe, and Northern North America; for the latitudes which in the north include these areas, south of the equator embrace little else than water. Darwin and others, however, have found in that part of South America situated within the temperate zone the traces of the last Glacial period, thus showing that its rigors were not confined to the northern hemisphere, but affected alike all parts of the earth's surface equally distant from the equator. It is our purpose, in the present chapter, to trace the consequences of the last Glacial period within the tropics, and see how far they may be conformable to the theory of the transverse rotation of the earth.

In studying the phenomena of the glacial interval of Pliocenic epoch, naturalists, very generally, have seemed to ignore the intercalated traces of warmth plainly indicated in the geological record, and to dwell upon and exaggerate the perhaps no more remarkable, although more abundant signs of cold. This course has led to the adoption of the theory of a "cosmic or geologic winter," according to which the temperature of the whole earth became largely reduced throughout an interval of time amounting to thousands of centuries; and the traces of warmth, whenever adverted to, were held to denote lesser intervals, during which the normal rigor of the "cosmic winter" became, in some inexplicable manner, somewhat modified. Among the foremost advocates of this theory was the late Professor Louis Agassiz,* whose familiarity with the phenomena of glaciation, both ancient and modern, cannot but invest his opinions, in questions of fact at least, with the rank of undoubted authority. With implicit faith in the hypothesis of a cosmic winter, and be-

* Professor Agassiz' death occurred while the present chapter was in the course of preparation.

holding in the impressions left by the cold of the severe winters of the Glacial period evidences of a degree of refrigeration such as had buried the continents of Europe and North America beneath superincumbent masses of ice from five to ten thousand feet in thickness, the natural sequence of such belief, namely, that such excessive cold could not have been confined to high and middle latitudes, but must have extended nearly, if not quite, to the equator, became firmly impressed upon his mind. Rest, and change of scene and climate, having become necessary to him on account of impaired health, he was fortunately successful, in the winter of 1865, in organizing an expedition for the purpose of scientific investigation, with the double object in view, first, to study the ichthyological fauna of Brazil in its native habitat — a desire prompted by the fact that, when a student of twenty years of age, he had, on the death of Spix, been intrusted by Martius with the work of describing the fishes which those two naturalists had brought back with them from their celebrated Brazilian journey; and, secondly, to search within the tropics for the traces, which he supposed must there exist, of that period of desolation and death his imagination had evoked from the monuments of the Glacial period. The result was, in the latter case, as we shall see, the gathering together of a mass of facts which, at first sight, appeared to confirm his theory, but to those who held his views to be untenable, seemed so monstrous and so out of the ordinary course of nature, that they felt impelled either to doubt their authenticity, or to place them in the rapidly accumulating category of irreduceable physical phenomena which has so long perplexed, baffled, and misled philosophers.

An interesting account of the expedition is given to the world in the work entitled "A Journey in Brazil;" and, believing the idea preposterous, that the professor, with his peculiar advantages and large experience as a practical observer of glacial phenomena could have been mistaken in assigning a glacial origin to the facts therein narrated, we cannot but accept them as evidences of former ice-action, although compelled to dissent from the conclusion derived therefrom relative to the excessive degree of cold assigned as the producing agency.

In mapping out the labor to be performed by him and his assistants in a series of lectures delivered on shipboard during the passage, the plan of geological investigation with special reference to glacial phenomena is laid down as follows: "The basin of the Amazons, for instance, is a level plain. The whole of it is covered with loose materials. We must watch carefully the character of these loose materials, and try to track them to their origin. As there are very characteristic rocks in various parts of this plain, we shall have a clew to the nature of at least some portion of these materials. My own previous studies have given me a special interest in certain questions connected with these facts. What power has ground up these loose materials? Are they the result of disintegration of the rock by ordinary atmospheric agents, or are they caused by the action of water, or by that of glaciers? Was there ever a time when large masses of ice descended far lower than the present snow line of the Andes, and, moving over the low lands, ground these materials to powder? We know that such an agency has been at work on the northern half of this hemisphere. We have now to look for its traces on the southern half, where no such investigations have ever been made within its warm latitudes; though to Darwin science is already indebted for much valuable information concerning the glacial phenomena of the temperate and colder portion of the South American continent." *

In a subsequent lecture upon the traces of glaciers as they exist in the northern hemisphere, and the signs of the same kind to be sought for in Brazil, he says, "When the polar half of both hemispheres was covered by such an ice shroud, the climate of the whole earth must have been different from what it is now. The limits of the ancient glaciers give us some estimate of this difference, though of course only an approximate one. A degree of temperature in the annual average of any given locality corresponds to a degree of latitude; that is, a degree of temperature is lost for every degree of latitude as we travel northward, or gained for every degree of latitude as we travel southward. In our times, the line at which

* A Journey in Brazil, p. 15.

the average annual temperature is 32°, that is, at which glaciers may be formed, is in latitude 60° or thereabouts, the latitude of Greenland; while the height at which they may originate in latitude 45° is about 6,000 feet. If it appear that the ancient southern limit of glaciers is in latitude 36°, we must admit that in those days the present climate of Greenland extended to that line. Such a change of climate with reference to latitude must have been attended by a corresponding change of climate with reference to altitude. Three degrees of temperature correspond to about one thousand feet of altitude." *

With these data, and taking the present line of perpetual snow under the equator at 15,000 feet, Professor Agassiz assumes to have the means, when he shall have discovered the limit of ancient glacial action in Brazil, of determining the mean annual temperature of the tropics in glacial times. Assuming, for illustration, 7,000 feet of altitude as the lower limit of ancient glaciation, he infers from it that the temperature was about 24° below that of the present time; but in an appended note he observes that "it proved in the sequel unnecessary to seek the glacial phenomena of tropical South America in its highest mountains. In Brazil the moraines are as distinct and as well preserved in some of the coast ranges on the Atlantic side, not more than twelve or fifteen hundred feet high, as in any glaciated localities known to geologists in more northern parts of the world. The snow line, even in those latitudes, then descended so low that masses of ice formed above its level actually forced their way down to the sea-coast."

On the Professor's arrival in Rio de Janeiro, his "attention was immediately attracted by a very peculiar formation consisting of an ochraceous, highly ferruginous, sandy clay." This deposit, afterwards found to extend throughout the entire Amazonian valley, proved on examination to be one of the forms of glacial drift so familiar to him. In a letter to Professor Peirce, dated at Tijuca, seven or eight miles from Rio, he writes that from the terrace of his hotel can be seen a "drift hill with innumerable erratic boulders, as characteristic as any I have ever seen in New England. I had before seen sundry unmistakable traces

* A Journey in Brazil, p. 18.

of drift, but there was everywhere connected with the drift itself such an amount of decomposed rocks of various kinds, that, though I could see the drift and distinguish it from the decomposed primary rocks in place, on account of my familiarity with that kind of deposits, yet I could probably never have satisfied anybody else that there is here an equivalent of the Northern drift, had I not found yesterday, near Bennett's hotel at Tijuca, the most palpable superposition of drift and decomposed rocks, with a distinct line of demarkation between the two, of which I shall secure a good photograph. This locality afforded me at once an opportunity of contrasting the decomposed rocks which form a characteristic feature of the whole country (as far as I have yet seen it) with the superincumbent drift, and of making myself familiar with the peculiarities of both deposits; so that I trust I shall be able hereafter to distinguish both, whether they are in contact with one another, or found separately." After a description of the formation composed of the disintegrated material of the rocks in place, he proceeds to say in reference to it, "that such masses forming everywhere the surface of the country should be a great obstacle to the study of the erratic phenomena is at once plain, and I do not therefore wonder that those who seem familiar with the country should now entertain the idea that the surface rocks are everywhere decomposed, and that there is no erratic formation or drift here. But upon close examination it is easy to perceive that, while the decomposed rocks consist of small particles of the primitive rocks which they represent, with their dikes and all other characteristic features, there is not a trace of larger or smaller boulders in them; while the superincumbent drift, consisting of a similar paste, does not show the slightest sign of the indistinct stratification characteristic of the decomposed metamorphic rocks below it, nor any of the decomposed dikes, but is full of various kinds of boulders of various dimensions. . . . But you see that I need not go to the Andes to find erratics, though it may yet be necessary for me to go, in order to trace the evidence of glacier action in the accumulation of this drift; for you will notice that I have only given you the evidence of extensive accumulations of drift similar in its characteristics to Northern drift. But I have not yet seen a trace of glacial action, properly speaking, if polished surfaces, and scratches, and furrows are especially to be considered as such.

"The decomposition of the surface rocks to the extent to which it takes place here, is very remarkable, and points to a new geological agency, thus far not discussed in our geological theories. It is obvious here (and to-day with the pouring rain which keeps me in doors I have satisfactory evidence of it) that the warm rains falling upon the heated soil must have a very powerful action in accelerating the decomposition of rocks. It is like torrents of hot water falling for ages in succession upon hot stones. Think of the effect, and, instead of wondering at the large amount of decomposed rocks which you meet everywhere, you will be surprised that there are any rocks left in their primitive condition." *

After a sojourn of two months in the country, and having overcome the difficulties at first encountered in tracing the erratic drift, and distinguishing it from the decomposed rocks in place, Professor Agassiz finds "no more difficulty in following the erratic phenomena in these Southern regions than in the Northern hemisphere. All that is wanting to complete the evidence of the actual presence of ice here, in former times, is the glacial writing, the striæ and furrows and polish which mark its track in the temperate zone. These one can hardly hope to find where the rock is of so perishable a character and its disintegration so rapid. But this much is certain, — a sheet of drift covers the country, composed of a homogeneous paste without trace of stratification, containing loose materials of all sorts and sizes, imbedded in it without reference to weight, large boulders, smaller stones, pebbles, and the like." †

On the voyage up the coast from Rio to Para, at the mouth of the River Amazon, the drift was followed, and carefully examined at every station. "At Bahia it contained fewer large boulders than in Rio, but was full of small pebbles, and rested upon undecomposed stratified rock. At Maceió, the capital of the province of Alagôas, it was the same, but resting upon decomposed rock, as at

* A Journey in Brazil, p. 86. † Ibid, p. 99.

Tijuca. Below this was a bed of stratified clay, containing small pebbles. In Pernambuco, on our drive to the great aqueduct, we followed it for the whole way; the same red clayey homogeneous paste, resting there on decomposed rock. The line of contact at Monteiro, the aqueduct station, was very clearly marked, however, by an intervening bed of pebbles. At Parahyba do Norte the same sheet of drift, but containing more and larger pebbles, rests above a decomposed sandstone somewhat resembling the decomposed rock of Pernambuco. . . . In the neighborhood of Cape St. Roque we came upon sand-dunes resembling those of Cape Cod, and wherever we sailed near enough to the shore to see the banks distinctly, as was frequently the case, the bed of drift below the shifting superficial sands above was distinctly noticeable. The difference in color between the white sand and the reddish soil beneath made it easy to perceive their relations. At Ceará, where we landed, Mr. Agassiz had an opportunity of satisfying himself of this by closer examination. At Maranham the drift is everywhere conspicuous, and at Pará equally so. This sheet of drift which he has thus followed from Rio de Janeiro to the mouth of the Amazons is everywhere of the same geological constitution. It is always a homogeneous clayey paste of a reddish color, containing quartz pebbles; and, whatever be the character of the rock in place, whether granite, sandstone, gneiss, or lime, the character of the drift never changes or partakes of that of the rocks with which it is in contact. This certainly proves that, whatever be its origin, it cannot be referred to the localities where it is now found, but must have been brought from a distance." * The presence of the drift was observed by the Professor's assistants in the environs of Barbacena and Ouro-Preto, and in the valley of the Rio das Velhas. Mr. Frederick C. Hartt, accompanied by Mr. Copeland, one of the volunteer aids of the expedition, who had been making collections and geological observations in the province of Spiritu Santo, in the valley of the Rio Doce, and afterwards in the valley of the Mucury, informed him "that he has found everywhere the same sheet of red, unstratified clay,

* A Journey in Brazil, p. 146.

with pebbles and occasional boulders overlying the rock in place;"* and Major Coutinho, a member of the Brazilian government corps of engineers, whom the Emperor had permitted to accompany and assist the Professor on the expedition, assured him, as soon as he became able to distinguish the drift from the decomposed rock, that it was to be found throughout the valley of the Amazon.

In the account of the voyage up the river Amazon, frequent mention is made of this deposit; and in every locality it is represented as presenting the same peculiar distinctive characteristics.

When in the vicinity of Ega, or Teffé, as the Brazilians call it, the opportunity there presented of prosecuting his ichthyological studies was so favorable, that Professor Agassiz with reluctance decided to forego his contemplated journey into Peru to visit at least the first spur of the Andes, with the purpose of ascertaining whether any vestiges of glaciers are to be found in the valleys — an expedition which, although the Professor was apprehensive that the warm torrential rains of these latitudes had decomposed the surfaces of the rocks, and obliterated all traces of glaciation, might have led to the discovery of the source whence proceeded the drift material deposited throughout the Amazonian valley. The drift is observed at Teffé, and "the more he considers . . . the more does he feel convinced that the whole mass of the reddish, homogeneous clay, which he has called drift, is the glacial deposit brought down from the Andes, and worked over by the melting of the ice which transported it."† At Obydos the drift deposit is described as being more full of pebbles than at Manaos, or Teffé, the same being deposited in lines or horizontal layers, such as were found in the same deposit along the coast and in the neighborhood of Rio; and its presence was also noted at the town of Viga, at the village of Sourés, and, in fact, in every locality where the geological aspect of the country is mentioned.

On the northern flank of the Serra of Ereré, one of a low range of hills on the north bank of the Amazon, not

* A Journey in Brazil, p. 404. † Ibid., p. 250.

far from it, and nearly parallel with its course, extending from the neighborhood of Almeyrim to the heights of Obydos, Professor Agassiz encountered "*genuine erratic boulders* . . . entirely distinct from the rock of the Serra, and consisting of masses of compact hornblende." The latitude of this locality is about 2° S.*

On the return voyage from Para to Rio de Janeiro, the expedition landed again at the town of Ceará, situated on the coast between Maranham and Cape St. Roque, and within four or five degrees of the equator, for the purpose of making a more careful examination of the geology of the coast; and also to afford Professor Agassiz an opportunity to satisfy himself by direct investigation as to the former existence of glaciers in the serras of this province. Having assumed, from the facts which we have cited in this chapter, that in the Glacial period, the Amazonian valley became filled with an enormous glacier, it was in this vicinity he expected to meet with traces of its southern lateral moraine. Dr. Felice, whose occupation as land-surveyor had familiarized him with the region of the Serra Grande, of which he had made a valuable map, and whose information was given with a degree of precision attesting its reliability, "tells Mr. Agassiz that there is a wall of loose materials, boulders, stones, &c., running from east to west for a distance of some sixty leagues from the Rio Aracaty-Assù to Bom Jesu, in the Serra Grande. From his account, this wall resembles greatly the 'Horse-backs' in Maine, those remarkable ridges accumulated by ancient glaciers, and running sometimes uninterruptedly for thirty or forty miles." † Unfortunately the state of the roads at that season of the year, and want of time, rendered a visit to this ridge impossible; and he was forced to content himself with an examination of the traces of local glaciers in the vicinity of Ceará.

On the way to Pacatuba, a village at the foot of the Serra of Aratanha, they followed a morainic soil for a great part of the journey, and passed many boulders on the road. An examination of the Serra of Monguba and its surroundings satisfied the Professor that "here too, all the valleys have had their glaciers, and that these valleys

* A Journey in Brazil, p. 418. † Ibid., p. 447.

(glaciers) have brought down from the hillsides into the plains boulders, pebble, and *débris* of all sorts." * The Serra of Aratanha was found, however, to furnish glacial phenomena in the greatest perfection, for there it is declared to be "as legible as in any of the valleys of Maine, or in those of the mountains of Cumberland in England. 'It had evidently a local glacier, formed by the meeting of two arms, which descended from two depressions spreading right and left on the upper part of the Serra, and joining below in the main valley. A large part of the medial moraine formed by the meeting of these two arms can still be traced in the central valley. One of the lateral moraines is perfectly preserved, the village road cutting through it; while the village itself is built just within the terminal moraine, which is thrown up in a long ridge in front of it." †

The important bearing of these facts upon our subject, showing as they do the prevalence of an intense degree of cold covering with snow and ice the sides of mountains of moderate elevation, situated within three or four degrees of latitude of the equator, warrants us in transcribing the Professor's detailed description of the same, which is as follows: "I spent the rest of the day in a special examination of the right lateral moraine, and part of the front moraine of the glacier of Pacatuba; my object was especially to ascertain whether what appeared a moraine at first might not, after all, be a spur of the Serra, decomposed in place. I ascended the ridge to its very origin, and there crossed into an adjoining depression, immediately below the Sitio of Captain Henriquez, where I found another glacier bottom of smaller dimensions, the ice of which probably never reached the plain. Everywhere in the ridges encircling these depressions the loose materials and large boulders are so accumulated and embedded in clay or sand, that their morainic character is unmistakable. Occasionally, where a ledge of the underlying rock crops out, in places where the drift has been removed by denudation, the difference between the moraine and the rock decomposed in place is recognized at once. It is equally easy to distinguish the boulders which here and there

* A Journey in Brazil, p. 454. † Ibid., p. 456.

have rolled down from the mountain, and stopped against the moraine. The three things are side by side, and might at first be easily confounded; but a little familiarity makes it easy to distinguish them. Where the lateral moraine turns towards the front of the ancient glacier, near the point at which the brook of Patacuba cuts through the former, and a little to the west of the brook, there are colossal boulders leaning against the moraine, from the summit of which they have probably rolled down. Near the cemetery the front moraine consists almost entirely of small quartz pebbles; there are, however, a few larger blocks among them. The medial moraine extends nearly through the centre of the village, while the left-hand lateral moraine lies outside of the village, at its eastern end, and is traversed by the road leading to Ceará. It is not impossible that eastward a third tributary of the Serra may have reached the main glacier of Pacatuba. I may say, that in the whole valley of Hasli there are no accumulations of morainic materials more characteristic than those I have found here, — not even about the Kirchet; neither are there any remains of the kind more striking about the valleys of Mount Desert in Maine, where the glacial phenomena are so remarkable, nor in the valleys of Lough Fine, Lough Augh, and Lough Long in Scotland, where the traces of ancient glaciers are so distinct. In none of these localities are the glacial phenomena more legible than in the Serra of Aratanha." *

On the return to Rio, the Organ Mountains in the vicinity of that city were visited, and were found to exhibit abundant evidence of ice-action in former times. Moraines, morainic soils, or accumulations of drift, with all sorts of loose material buried in it, together with erratic boulders, entirely distinct from the rock in place, were found in abundance all over that region; but the heavy growth of forest, by covering the inequalities of the soil, made the study of glacial phenomena difficult, and but few details are given.

The foregoing citations embrace, substantially, the results of Professor Agassiz' search for glacial phenomena in tropical South America. He and his assistants, as far

* A Journey in Brazil, p. 463.

as their observations extended, found the whole surface of the country from Rio de Janeiro to the River Amazon covered with a deposit which, from his description, must be classed as genuine glacial drift. Erratic boulders were seen in many places, and although not frequent near the great river, occurred at one place in its immediate vicinity, within about two degrees of latitude of the equator. In nearly the same latitude occur the not more remarkable, but, if possible, more positive and unequivocal traces of ice-action in the remains of the glaciers of the Serras of Monguba and Aratanha, where the glacial phenomena are declared to be as distinct and legible as in any of those localities in the northern hemisphere, which have become celebrated as affording the most favorable opportunity for its observation. Now, is this evidence reliable? Even if it is admitted, as has been suggested, that the deposit Professor Agassiz has classed as glacial drift, is in some places loéss, or inundation-mud, and in others the produce of land floods,* can there be any mistake as to the identity of the erratics of Ereré and other places, and the glaciers of Pacatuba and of the Organ Mountains? Could this distinguished naturalist and philosopher, one of whose chief pursuits has been to study the phenomena of glaciation, ancient and modern, in all its forms, have deceived himself in this instance? Could it be possible for him to do so, when the facts were so palpable and clear? The idea is preposterous and absurd. Whatever notions the results of his observations may conflict with, whatever hypotheses they may overturn, the fact of ancient glacial action in the tropics must be accepted as an undoubted fact.

If, then, glacial processes were in progress at the equator during the Glacial period, under the hypothesis that the stability of nature directly depends on the preservation of approximating the present inclination of the terrestrial axis, involving the present relative position of the earth and sun, the theory of a cosmic winter consequent on a diminution of the solar influence becomes inevitable; for such diminution is the only rational cause conceivable which could produce an arctic climate under the sun's vertical rays. If, therefore, under the present status, we

* Prin. of Geol., vol. i. p. 464.

find from the foregoing facts, that a degree of cold prevailed at the equator, within the Glacial epoch, equal to 32° F., or that required for the formation of glaciers, such glaciers descending to the plains from mountains of moderate elevation and reaching the sea, the heating effects of the sun's rays must have been but a small fraction of what they are now; or, in other words, the temperature of the sun must have been reduced in such a degree that its vertical rays, at the equator, would there produce only the same thermal effect they now do under an inclination from the vertical of 65° or 70° in Greenland. With an arctic climate or a mean annual temperature of 32° F., or less, at the equator, what must have been the climatal condition of other portions of the surface of the earth? They must certainly have experienced a degree of cold sufficient to exterminate all forms of life, arrest all meteorological processes, and solidify all the aqueous part within a very short space of time, leaving only a dreary blank to mark a long period in the history of the world.

It does, indeed, seem impossible for any candid mind to contemplate the globe, revolving throughout a vast cycle of ages, as a ball of ice, deprived in a large measure of the light and heat of the sun, such deprivation entailing inevitable extinction to all the forms of life inhabiting it, and at the termination of such cycle, necessitating a general re-creation, without a fervent protest against so irrational and improbable a system — one in such direct antagonism to all the modes of action exhibited in the usual operations of nature.

The fact of tropical glaciation, taken in connection with the hypothesis which assumes the present inclination of the earth's axis to be, approximately, a constant quantity, involves, as we see, a large amount of solar variability. Now, it is perfectly natural to suppose that, did such variation really occur, covering periods of time equal to thousands of centuries, they would be brought about in a uniform and progressive manner; that is, assuming the present as the usual mean of temperature, the change to a greater or less degree, and the return to the normal condition, would proceed uninterruptedly and uniformly from the normal or mean point to the extremes, and conversely from the extremes to the mean. The weight of evidence, however, is largely against this supposition, for we have adverted to Lyell's intercalated seasons of warmth of the Ice period, also to traces of glacial action in seasons of supposed abnormally high temperature; and Professor Agassiz, in treating of the northern and southern "ice-caps," which he supposed covered nearly the whole of the solid portion of the earth's surface in glacial times, speaks of the alternate freezing and thawing of the ice and snow. These facts, under the hypothesis that assumes the permanence of the present inclination of the earth's axis, tend to show that the sun, which we are naturally inclined to regard as one of the most stable objects in the universe, is liable, at irregular intervals, to exhibit excessive degrees of variability; such changes, in their progress, being subject to the most capricious fluctuations, making the sun, on the whole, instead of the most constant, one of the most fickle objects in nature.

The argument already advanced as showing the absurdity of supposing that the present climatic relations between polar and equatorial latitudes could have prevailed during the eras of polar warmth, is inversely applicable in the present instance. In the former case, assuming the inclination of the terrestrial axis to approximate its present value, all organic forms inhabiting the earth, excepting only those living within the polar circles, must inevitably have been destroyed by the intense heat of the sun. In the latter, the agency of cold must have entailed a similar destruction of life in all but equatorial localities. The reader is earnestly requested to observe that these are not idle speculations, based on loose and indefinite data, but are infallible deductions from well-ascertained facts, and well-known inherent properties of matter.

But we have not yet reached the climax of absurdity to which we must arrive under the hypothesis that the stability of nature depends on the maintenance of the present inclination of the terrestrial axis of rotation. Even if we can so far do violence to our reasoning faculties as to admit that, under the present inclination of the sun's rays as seen at the poles, a luxuriant vegetation and a varied and abundant animal life flourished there within the Glacial period, and also, on the other hand, that under the direct influence of the vertical rays of the sun, as now seen at the equator, natural processes of glaciation could there proceed on a scale of more or less magnitude, these are not the ultimate tests of our credulity; for we must still further admit that these conditions of heat at the poles, and refrigeration at the equator, could have been exhibited at one and the same time, — the synchronism of glacial deposits all over the world being undoubted. In other words, we must believe that, at the same time when the heating effect of the rays of the sun had become so reduced as to be insufficient to prevent the formation of glaciers at moderate elevations within the tropics, the same rays impinging the earth's surface in polar regions, at the angle they now do, produced heating effects so marked as at first to give the impression of the prevalence of a purely tropical climate. From this dilemma there is no escape.

It is beyond question that, during a certain period of the world's history, a comparatively cold climate at the equator coexisted with a comparatively warm climate at the poles. Under the present status, we see that such conditions are absolutely impossible. There remains but one rational method to account for the facts; and that is, to suppose that the whole annual amount of the solar influence enjoyed by the earth was in those times more evenly distributed over its surface than now. There appears to be but one way by which changes in the mode of distribution of the solar light and heat can be effected, and that is by changes in the mode of presentation of the earth's surface to the sun. The annual and diurnal motions continuing the same, the only other terrestrial motion possible, involving change in the mode of presentation of the earth's surface to the sun, is a movement of the poles at a right angle with the equator, transverse to the axis of diurnal rotation. Such a motion we know is now in progress; and we have but to look upon it as a continuous, instead of an oscillatory one, as has been heretofore supposed, in order to find a simple and rational escape from all our difficulties.

If, then, this motion is continuous, involving complete revolutions of the earth, there are times during each rotation when the polar axis of rotation becomes coincident with the orbital plane. This position of the earth has been before alluded to in connection with the evidences of warmth at the poles in former times. Now, if the annual amount of heat received by the earth from the sun is continually the same, whenever any portion of its surface shall receive a larger proportional share of the same than it now does, other portions must receive proportionally less; and this is exactly what our facts indicate. While polar areas were enjoying a more than torrid heat for half of the year, equatorial regions suffered a corresponding loss; and hence these evidences of cold within the tropics.

The single day and night at the poles, resulting from this position of the earth, representing respectively summer and winter, would, at the equator, be contrasted by a highly diversified compound arrangement of seasons.

The annual course of the sun north and south, instead of being confined as at present within the limits of 23° 28′ each side of the equator, would extend from one celestial pole to the other, reaching the northern and southern horizons, respectively, at intervals of six months. The length of the days would, therefore, vary from twelve hours to nothing, and the length of the nights from twelve to twenty-four hours, in the same time that the angle of inclination of the sun's rays would vary from perpendicularity to ninety degrees, or coincidence with the horizon. From this arrangement would result two summers and two winters in each year; sudden transitions from heat to cold and from cold to heat, which would be likely, under varying circumstances, to give rise to extraordinary climatal peculiarities.

Allowing that the ordinary heat-distributing agencies exhibited at that time their present activity (although they may well be supposed more efficacious), they, in connection with those other local causes affecting the climates of particular localities, must have produced great contrasts of temperature in places not, perhaps, far removed from each other. In places favorably situated these agencies may be imagined capable of preventing the heat of the summers from becoming greatly reduced during the short winters, while in others opposite causes would operate to neutralize the effects of the summer heats, so that we might be justified in expecting, under the circumstances, to find glacial phenomena in progress in immediate juxtaposition with a continuous growth of *modified* tropical vegetable forms. Whether the mean annual temperature there became so reduced that glaciers descended from the mountains and invaded the plains, is a question which does not affect the truth of our theory. Were such indeed the case, it might perhaps make it more difficult to perceive how tropical forms could have been carried through such an epoch. There is no question but that an abrupt transition from the present climate of the tropics to the one our most probable conjectures would assign as that of the Glacial period, or time of axial coincidence with the ecliptic, would be largely destructive to present organic forms. But if we suppose such change brought about in so slow and uniform a manner that nearly a million of years was required for its completion, although it might induce a large amount of modification and some extinction (if the supplanting of old forms by their improved descendants can properly be so termed), we can hardly expect to encounter throughout the whole period any point in the same marked by an extensive destruction of life. To repeat a former illustration, the changes in the organic world and its surrounding conditions would be so gradual that an intelligent being, endowed with life, and inhabiting the earth for a period of ten thousand years, would observe no change in the prevailing status, several such cycles being required to make its progress manifest. Thus, what at first sight appears to be a serious objection, — the difficulty of carrying tropical forms through a period of comparative cold, — when viewed in this light ceases to bear the aspect of antagonism to our system; and it can with truth be affirmed that, as far as present observations extend, the whole arcana of nature, while furnishing facts in abundance confirmatory of its truth, contain not one that can be said to be in serious conflict with it.

CHAPTER V.

A BRIEF REVIEW OF VARIOUS HYPOTHESES ADVANCED TO ACCOUNT FOR THE CLIMATAL VICISSITUDES OF FORMER TIMES.

LYELL ON MR. EVANS'S SUGGESTION OF A CHANGE IN DIRECTION OF THE AXIS OF THE EARTH'S CRUST. — ON M. POISSON'S HYPOTHESIS. — LYELL'S THEORY OF GEOGRAPHICAL CHANGE.

IF the evidence of great changes of climate upon the earth's surface, since it first became the habitation of living beings, may be considered as conclusive, the next inquiry is, To what agency have these changes been referred by those who have given their attention to the subject?

It may first be stated that these changes, taken as they usually are, in connection with the idea of the permanence of the present relative position of the earth and sun, have presented to physicists, borrowing their own phraseology, some of the most perplexing problems in nature. Being clearly antagonistic, and utterly irreconcilable, the attempt to account for former vicissitudes of climate, under the existing order of things, has led to theories of the most extravagant character, the only result, perhaps, to be reasonably looked for in such a case.

"The earlier speculators in geology," says Lyell, "availed themselves of this, as of every obscure problem, to confirm their views concerning a period when the planet was in a nascent or half-formed state, or when the laws of the animate and inanimate world differed essentially from those now established; and in this, as in many other cases, they succeeded, to no small extent, in diverting attention from that class of facts which, if fully understood, might have led the way to an explanation of the phenomena. At first it was imagined that the earth's axis had been for ages perpendicular to the plane of the ecliptic, so that there was a perpetual equinox, and uniformity of seasons throughout the year; that the planet enjoyed this 'paradisiacal' state until the era of the great flood; but in that catastrophe, whether by the shock of a comet, or some other convulsion, it lost its equipoise, and hence the obliquity of its axis, and with that the varied seasons of the temperate zone, and the long nights and days of the polar circles.

"When the progress of astronomical science had exploded this theory, it was assumed, that the earth at its creation was in a state of igneous fluidity, and that, ever since that era, it had been cooling down, contracting its dimensions, and acquiring a solid crust. It was also taken for granted that this original crust was the same as that which we are now studying, and which contains the monuments of a long series of revolutions in the animate world. This notion, however arbitrary, was well calculated for lasting popularity, because it referred the mind directly to the beginning of things, and required no support from any ulterior hypothesis. But the progress of geological investigation gradually dissipated the idea, at first universally entertained, that the granite or crystalline foundations of the earth's crust were of older date than all the fossiliferous strata. It has now been demonstrated that this opinion is so far from the truth, that it is difficult to point to a single mass of volcanic or plutonic rock which is more ancient than the oldest known organic remains. Such being the case, the question of original fluidity, although a matter of legitimate speculation to the physicist, is one with which the geologist is but little concerned. It may relate to a state of things, which preceded our earliest records by a lapse of ages many times greater than the entire series of geological epochs with which we are acquainted." *

Instead of continuing to follow our author as he proceeds to elaborate his own theory, we will first notice some of those which may, perhaps, be considered of minor importance.

I am not aware that the attempt has ever been made to account for the climatal changes of former times on the

* Prin. of Geol., vol. i. p. 233.

direct assumption of solar variability, although that idea would seem to underlie the hypothesis of a "cosmic winter." Questions concerning the solar constitution, bearing upon this point, will be considered in a subsequent chapter.

Mr. Evans, in the year 1866, according to Lyell, suggested that former changes of climate might be connected with the sliding of a solid shell, forming the crust of the earth, over an internal fluid nucleus. It was supposed that the equilibrium of the external shell might have been "disturbed by the transfer of the sediment from one part of the surface to another, or by the upheaval of new continents and islands; and Mr. Evans shows that, whenever matter is abstracted from one part and added to another, the centrifugal force of the augmented extraneous matter would tend to draw over the shell towards the equator, or an opposite effect would be produced if the surface was relieved of part of its weight, in which case the lighter part would move towards the pole."* Were such an operation possible, a given portion of the terrestrial surface, under the present status, might alternately be subjected to a torrid, a temperate, or an arctic climate.

To this it may be said, in the first place, that there are no sufficiently valid reasons to justify the current belief in the doctrine which considers the earth a mass of intensely heated molten matter, with an exterior crust or shell of comparatively little thickness — a belief evidently based on the fact that as we descend into the earth, its temperatu e is observed to rise in a ratio proportional to the depth attained, and the assumption, by no means warranted by the fact, that this increase of heat must continue to the earth's centre.

If the world was originally an incandescent fluid mass, and has assumed its present aspect by radiating into space the excess of heat, there is no reason to suppose that the cooling process proceeded under other principles than those governing the cooling off of smaller fluid masses. The transformation of a fluid world to a solid one, would be effected in precisely the same way as a quantity of molten metal in a foundery becomes solidified, or as water in a bucket is converted into ice. The whole must be-

* Prin. of Geol., vol. ii. p. 208.

come reduced in temperature to a certain point before crystallization can begin, and we may be assured that if any very large portion of the material of the earth still remains in a molten condition, its temperature is not much, if any, in excess of that of the lavas ejected from the craters of volcanoes.

An attempt has been made by Mr. Hopkins to determine the least thickness that can be assigned to this supposed exterior crust. "This result," says Lyell, "he has endeavored to obtain by a new solution of the delicate problem of the precessional motion of the pole of the earth, caused . . . by the attraction of the sun and moon, and principally the moon, or the protuberant parts at the earth's equator; for if these parts were solid to a great depth, the motion thus produced would differ considerably from that which would exist if they were perfectly fluid, and incrusted over with a thin shell only a few miles thick. . . . Mr. Hopkins has, therefore, calculated the amount of precessional motion which would result if we assume the earth to be constituted as above stated; *i. e.*, fluid internally, and enveloped by a solid shell; and he finds that the amount will not agree with the observed motion, unless the crust of the earth be of a certain thickness." After making every allowance for certain doubtful elements, Mr. Hopkins's researches conducted him to the conclusion 'that the minimum thickness of the crust of the globe, which can be deemed consistent with the observed amount of precession, cannot be less than one-fourth or one-fifth of the earth's radius;' that is, from eight hundred to one thousand miles; and Lyell remarks, that "this is a *minimum*, and any still *greater* amount would be quite consistent with the actual phenomena; the calculations not being opposed to the supposition of the general solidity of the entire globe." *

Waiving the above fatal objection to Mr. Evans's hypothesis, Lyell proceeds to examine it in connection with the idea of central fluidity. The arguments of Newton and Laplace against the probability of a shifting of the earth's axis of rotation are cited, which, however irrelevant in connection with the theory of a gradual motion, may still be pertinent and just in relation to sudden and violent

* Prin. of Geol., vol. ii. p. 203.

changes, consequent upon causes of the nature of that under consideration; and also that of Mr. Airy, who had "pointed out that the elevation of mountain chains at certain geological periods, which had been proposed as causing an alteration in the earth's centre of gravity, was an insignificant cause, since the size of such mountain masses was very minute, when compared to the equatorial protuberance." These arguments seem to relate to the shifting of the axis of the entire mass of the planet, while the suggestion of Mr. Evans is "that the axis of rotation of the nucleus might remain unchanged, while a solid shell, not more, perhaps, than twenty-five miles in thickness, might have its axis of rotation altered. To this hypothesis there are several objections: —

"First, in all geological times, the transfer of sediment has been taking place not only from higher to lower latitudes, but also from lower to higher. There is the like tendency in the various elevations and depressions of land simultaneously in progress to balance each other. It is only the excess of alteration in one direction that can be available as a disturbing cause, and we can hardly imagine this excess to be important enough to cause a sensible change in the axis of rotation even of the external shell, such as might explain the altered climate of the same country in successive geological periods.

"Secondly, a greater difficulty arises out of the fact that the earth is a spheroid, and not a perfect sphere, since it becomes necessary to imagine the fluidity of the nucleus to be so perfect as to allow the shell to slide freely over it. If the lower or inner surface of the envelope be irregular in shape, or if it be even viscous in part, great resistance would be offered to any change in its position. Its freedom of motion would be checked by its not fitting the nucleus, let its change of position be ever so slight, and this change could only be effected by the most violent friction, attended by the bending and rending of the incumbent mass." *

This hypothesis, which, were its substantiation possible, might, perhaps, be made to partially account for the climatal changes of former times, is thus shown to be a physical impossibility. Were it otherwise, as long as it

* Prin. of Geol., vol. ii. p. 209.

contemplates sudden and violent changes, grand catastrophes, the traces of which would everywhere abound, and be readily recognizable, and which, by suddenly precipitating new conditions of life, must have been largely fatal to all classes of organisms, who would lack the time necessary to effectuate modifications tending to adapt them to their changed surroundings, — these considerations alone justify us in dismissing it at once, as clearly opposed to all the analogies of nature.

In the year 1837, M. Poisson, a distinguished mathematician and philosopher, in endeavoring to account for secular variations in climate, suggested that they might be due to the passage of the solar system through regions of space varying in temperature, from experiencing a greater or less degree of stellar light and heat, according to the number and proximity of the stellar bodies to such regions. "He begins," says Lyell, "by assuming, first, that the sun and our planetary system are not stationary, but carried onward by a common movement through space. Secondly, that every point in space receives heat as well as light from innumerable stars surrounding it on all sides, so that if a right line of indefinite length be produced in any direction from such point, it must encounter a star either visible or invisible to us. Thirdly, he then goes on to assume, that the different regions of space, which in the course of millions of years are traversed by our system, must be of very unequal temperature, inasmuch as some of these must receive a greater, others a less quantity of radiant heat from the great stellar enclosure. If the earth, he continues, or any other large body, pass from a hotter to a colder region, it would not readily lose in the second all the heat which it has imbibed in the first region, but retain a temperature increasing downwards from the surface, as is the actual condition of our planet.

"Now the opinion originally suggested by Sir W. Herschel, that our sun and its attendant planets were all moving onward through space, in the direction of the constellation Hercules, is very generally thought by modern astronomers to be confirmed. But the amount of the movement is still uncertain, and great indeed must be its extent before this cause alone can work any material alteration in the terrestrial climates. Mr. Hopkins, when treating of this theory, remarked that so far as we are acquainted with the position of stars not very remote from the sun, they seem to be so distant from each other, that there are no points in space among them where the intensity of radiating heat would be comparable to that which the earth derives from the sun, except at points very near to each star. Thus, in order that the earth should derive a degree of heat from stellar radiation comparable to that now derived from the sun, it must be in close proximity to some particular star, leaving the aggregate effect of radiation from the other stars nearly the same as at present. This approximation, however, to a single star could not take place consistently with the preservation of the motion of the earth about the sun, according to its present laws.

"Suppose our sun should approach a star within the present distance of Neptune. That planet could no longer remain a member of the solar system, and the motions of the other planets would be disturbed in a degree which no one has ever contemplated as probable since the existence of the solar system. But such a star, supposing it to be no larger than the sun, and to emit the same quantity of heat, would not send to the earth much, more than one-thousandth part of the heat which she derives from the sun, and would therefore produce only a very small change in terrestrial temperature." *

Lyell's Theory. — In the endeavor to reconcile former vicissitudes of climate with the existing terrestrial status, Lyell, who tacitly ignores the theory of "geologic" winters and summers, involving large amounts of change in the mean temperature of the earth, and who seems disposed to view each and every instance of departure from the prevailing climatic conditions of the present time as the chance product of a separate and distinct combination of the effects of a class of secondary agencies of a more or less local character, no one of which is supposed of sufficient importance, of itself, to effect the results in question, enumerates as these agents, variations in the eccentricity of the earth's orbit, entailing change in the aphelion and perihelion distances of the earth from the sun; the precession of the equinoxes, combined with the revolution of the

* Prin. of Geol., vol. i. p. 295.

apsides; and lastly, geographical changes involving variation in the proportion of land and sea, and the elevation and contour of the land. Thus, for instance, where geology shows the prevalence of a warm climate in any part of the northern polar regions, it is assumed that certain possible geographical changes of the kind above indicated, combined with a decreased perihelion distance, the same occurring when precession had brought the northern summer coincident with it, have been the cause of such warm polar climate, and, on the other hand, that the appearances denoting the prevalence of a glacial period in the northern hemisphere are the result of an increased aphelion distance happening at a period of northern winters in aphelion, intensifying the effects of supposed geographical mutations.

The universal synchronism of glacial phenomena all over the earth affords good ground for a general denial of this hypothesis.

Variations in Eccentricity. — Of the assumed astronomical causes of change, variations in eccentricity are held to be of the most importance. The mean distance of the earth from the sun having been demonstrated by geometers to be invariable, it would naturally be supposed that the mean annual amount of light and heat received by the earth from the sun would also continue the same; but we are assured, by Sir John Herschel, that such is not the case, and that the whole amount of heat received in one revolution is inversely proportional to the minor axis of its orbit. The eccentricity of orbit being variable, involving changes in the value of the minor axis, a variation in the quantity of light and heat transmitted from the sun to the earth is supposed to ensue. But, according to the same author, as cited by Lyell, "As the extreme amount of difference in the quantity of heat annually received, owing to such change in the minor axis, can never by possibility exceed the whole supply in a ratio of more than 1,003 to 1,000, it may be neglected" in geological speculation of this nature.*

There is another way, theoretically at least, in which changes in eccentricity may affect climate. The heating

* Prin. of Geol., vol. i. p. 273.

effect of the solar rays is known to diminish in a ratio proportional to the distance they may have traversed from their source. The path of the earth about the sun being elliptical in form, and the sun's position with reference to ellipse being in one of the foci, it follows that the earth, in some parts of its annual revolution, must be further from the sun than in others. While the mean distance from the earth to the sun is ninety-one million four hundred thousand miles, the difference, at the present time, between the greatest or *aphelion* distance, and the least or *perihelion* distance, is about three millions miles ; that is, the earth, in its annual course, approaches this number of miles nearer the sun in December, during the northern winter, than it does in June, in the northern summer.

But this difference of three millions of miles, which now expresses the eccentricity of the terrestrial orbit, is not constant, but is subject to a continual variation from the attraction of the nearer and larger planets; Jupiter and Saturn exerting the principal influence in producing the perturbation. The calculations of Lagrange and Leverrier have shown that the minimum of eccentricity the earth's orbit can ever attain is about one half million of miles — a point it is now approaching, and which it will reach in about 23,900 years from the present time, becoming then as nearly circular in form as is possible; and that it may vary from this amount to any number within fixed limits, the greatest range of difference it can ever reach being fourteen million of miles. Thus, while the mean distance between the earth and sun remains invariable, the difference between perihelion and aphelion distance may vary from about one half million to fourteen millions of miles.

" Whatever be the ellipticity of the earth's orbit, says Sir John Herschel," as quoted by Lyell, " the two hemispheres must receive equal absolute quantities of light and heat per annum, the proximity of the sun in perigee or its distance in apogee exactly compensating the effect of its swifter or slower motion. But the same writer," Lyell continues, " in 1858, alluding to some speculations of Reynauld, speaks of the marked effects on climate which great variations in eccentricity might produce, causing the characters of the seasons in the two hemispheres to be strongly contrasted. So long as the position of the earth's perihelion remained the same as now, 'we should have in the northern a short but very mild winter, with a long but very cool summer — i. e., an approach to perpetual spring; while the southern hemisphere would be inconvenienced, and might be rendered uninhabitable by the fierce extremes caused by concentrating half the annual supply of heat into a summer of very short duration, and spreading the other half over a long dreary winter, sharpened to an intolerable intensity of frost when at its climax, by the much greater remoteness of the sun;' and he goes on to observe that, in consequence of the precession of the equinoxes, combined with the secular movement of the aphelion, the state of the northern and southern hemispheres here alluded to, would in the course of about 11,000 years be reversed, and as such alternations of climate must in the immense periods of the past which the geologist contemplates have happened, not once only, but thousands of times, 'it is not impossible,' he adds, 'that some of the indications of widely different climates in former times may be referable, in part at least, to this cause.'" *

If variations in eccentricity have materially aided geographical causes in producing the indications referred to, the effect, as above stated, would be to produce a decided difference in the climatic conditions of the northern and southern hemispheres. If the glacial period of the northern half of the globe resulted from geographical causes combined with the conjoint influence of winters happening in aphelion during an epoch of large eccentricity, even on the extremely improbable supposition that analogous physical processes were, at the same time, in operation in the southern, the decreased perihelion distance happening in summer would produce, in the last-mentioned hemisphere, a very equable, and, apparently, a very warm climate; the quality of equableness obviously tending to enhance the appearance of warmth. But if, as a more probable contingency, we suppose geographical causes inoperative, or acting in opposition south of the equator, the difference in climatic results must be still more pronounced, so that under the hypothesis we should have a glacial period in the northern hemisphere occurring at the same time with

* Prin. of Geol., vol. i. p. 274.

an era characterized by a great degree of warmth in the southern.

Notwithstanding the fact that an eminent naturalist (Mr. Darwin), in order to provide a way in which animal and vegetable life might have been preserved through the rigors of a cosmic winter, has, according to Lyell, inclined towards adopting Mr. Croll's theory of alternate glaciation and perpetual spring in the opposite hemispheres, on the ground that it would account for some anomalies in their distribution by affording a refuge for tropical life during a period of extreme cold, no geological fact appears more evident than that the peculiar conditions of the Glacial epoch pervaded alike all parts of the world, north and south, east and west, producing everywhere, according to latitude, precisely the same effects both in the organic and inorganic world. The drift deposits north of the equator appear to be identical and contemporary with those south of it; and the bulk-producing quality, whatever it may have been, of the conditions of life which prevailed during the drift period, is shown in as pronounced and as remarkable a degree in the herbivorous mammals of one hemisphere, as in those of the other. Were it possible, therefore, to show that geographical change, variations in eccentricity, and precession, assuming them all to be efficient agents, operated in conjunction at this period, as affecting climate, the result would not meet the requirements of the case, such results being of a local character, while the contemporaneous monuments of the Glacial period pervade alike every portion of the earth's surface.

In following Lyell through the elaborate argument contained in the thirteenth chapter of his Principles of Geology, relating to supposed astronomical causes affecting climate, one can hardly fail to note the want of correspondence between the results of theoretical deductions and those of actual observation. Abstractly, the truth of the proposition that the effect of the solar influence diminishes inversely as the squares of the distance traversed increase is undoubted; and it might be reasoned that when the earth is in that part of its orbit nearest the sun, it ought to receive a larger and when furthest a smaller amount of heat; and the difference in distance between perihelion and aphelion being no less than one-thirtieth of the mean distance, the planet ought therefore to be colder at one time and hotter at another, not merely by one-thirtieth of the heat received from the sun, but by one-fifteenth. And yet it is found by observation that the whole surface of the planet is actually warmer in June, when furthest, than in December, when nearest the sun. Again, Sir John Herschel "computes on theoretical grounds that there ought to be a difference of 23° F. when two places are compared at the same season and in the same latitudes on opposite sides of the equator; that is to say, the summer coinciding with perihelion ought to have a temperature of 11½° higher, and the winter in aphelion a temperature lower by the same amount, than the same seasons in the opposite hemisphere, where these astronomical conditions are reversed. The results of observation are not in harmony with this theory, the difference really indicated by the thermometer being only half that required by theory;" * and even this amount of difference is ascribed to the deficiency of land in the southern hemisphere, rather than to the astronomical cause under consideration, which is pronounced by the author to be "obviously insignificant."

There can be no doubt that, in the absence of the earth's atmospheric envelope, the difference between perihelion and aphelion heat would be immediately perceptible, and the whole surface of the earth warmer by one-fifteenth in December than in June. This envelope, however, seems to constitute a medium through which the perihelion excess of heat is arrested, and held in abeyance for a time, being permitted to become sensible at the surface of the earth only after an interval of six months, or not until the earth has arrived in that part of its orbit furthest from the sun. In the present state of our knowledge, who will venture to assert that the agency by which this compensation is affected is not adequate to distribute, in like manner, any possible excess of perihelion heat within the whole range of orbital eccentricity? Compensations analogous to this, it may be added, everywhere abound in nature.

In the same chapter an attempt is made to establish a correspondence between the occurrence of a large eccentricity and the date of the Glacial period, in order to locate,

* Prin. of Geol., vol. i. p. 282.

in time, the probable date of that epoch. It is not contended that great ellipticity of the earth's orbit exerted a controlling influence in determining the cold, but was only subordinate and auxiliary to the great primary cause, namely, geographical change.

In order to determine the extent to which changes in eccentricity might, in the past, have influenced climate, Mr. Stone of the Greenwich Observatory undertook, "by the use of Leverrier's formula, to determine when the last high eccentricity occurred. He found that it happened 210,065 years ago;" and Mr. Croll, following up Mr. Stone's calculations, computed the changes of eccentricity for a million years preceding and a million years following A. D. 1800. From the above, and other data, we have, on page 285 of the "Principles," a table, a portion of which is inserted on the next page, showing the variations that have taken place in the eccentricity of the earth's orbit for a million years previous to the year 1800.

A glance at this table is amply sufficient to show the remarkable degree of irregularity characterizing this perturbation; a consideration which alone is enough to indicate with certainty that no relation whatever can have existed between such perturbation and the climatal changes in question, the geological record unequivocally testifying to the uniform regularity with which those changes proceed. If we take the interval of 200,000 years embraced between the dates 700,000 and 900,000 years, in the table, we find the eccentricity at the latter date to be one and a quarter millions of miles, or five-twelfths of the present value, having decreased in the preceding 50,000 years from nine and a quarter millions to that number, which is the smallest indicated in the table. From this eccentricity of one and a quarter millions, the next 500 centuries show an increase to thirteen and a half millions, not only the highest number in the table, relating to eccentricity, but within half a million of the greatest value it can ever attain. In the next 500 centuries it decreases from thirteen and a half to two and a quarter millions; in the next, again increasing to ten and a half millions; while the next equal interval shows a decrease from the latter amount to four millions. Estimated by ordinary standards of comparison, fifty thousand years is indeed a great lapse of time; but, in a geological sense, it is small when contrasted with the vast epochs which constitute geological periods, or even the glacial intervals of such periods. In this light, a change from one set of climatal conditions to another, constituting their greatest possible contrast, within a period of 500 centuries, must be regarded as a decidedly sudden one. Again, it will be seen by consulting the table, that in the interval between 700,000 and 1,000,000 years ago, each included period of 100,000 years must have embraced both a glacial period and one of comparatively great warmth, making, in that time, three glacial, and, inclusive of the starting-point, four

Table showing the Variations in the Eccentricity of the Earth's Orbit for a Million Years before A. D. 1800.

	1.	2.	3.
	Number of years before A. D. 1800.	Eccentricity of orbit.	Difference of distance in millions of miles.
	0	.0168	3
	50,000	.0131	2¼
A	100,000	.0473	8½
	150,000	.0332	6
B { *a*	200,000	.0567	10¼
B { *b*	210,000	.0575	10½
	250,000	.0258	4½
	300,000	.0424	7¾
	350,000	.0195	3½
	400,000	.0170	3
	450,000	.0308	5½
	500,000	.0388	7
	550,000	.0166	3
	600,000	.0417	7½
	650,000	.0226	4
	700,000	.0220	4
C { *a*	750,000	.0575	10½
C { *b*	800,000	.0132	2¼
C { *c*	850,000	.0747	13½
	900,000	.0102	1¼
D	950,000	.0517	9¼
	1,000,000	.0151	2¾

COLUMN 1. Division of a million years preceding 1800 into twenty equal parts.

COLUMN 2. Gives the eccentricity of the earth's orbit in parts of a unit equal to the mean distance or half the longer diameter of the ellipse.

COLUMN 3. Gives in millions of miles the difference between the greatest and least distances of the earth from the sun, during the eccentricities given in column 2.

intervals of warmth. Geologists now seem to be unanimous in the opinion that the Ice period must have extended over hundreds of thousands of years, and our method of computation shows that nearly a million years were required to carry it through all of its phases, and thus the absurdity of crowding three of them, with their intercalated periods of warmth or repose, into a term of 3,000 centuries, must be sufficiently obvious.

Lyell informs us that, after mature deliberation, he has settled upon the period marked B., comprehended between the dates 210,000 and 200,000 years ago, as the probable date of the greatest cold. To the hypothesis which seeks to limit the time necessary to effect the physical and organic changes of the Ice period, or even the time of greater cold, to a period of 10,000 years, we append the author's own statement relative to the Glacial epoch, that not tens but hundreds of thousands of years would be required "for the changes in physical geography and organic life of which we have evidence." *

The conclusion then seems to be justified, that such comparatively rapid and highly irregular changes in eccentricity as are here indicated are clearly incompatible with the idea of their being in any degree whatever responsible for the graduated and uniform climatal changes of the geological record ; and we are warranted in the belief that not only must the amount, if any, of primary climatic variation resulting from different degrees of eccentricity, be very small, but also that the effect of variations in the same upon the annual round of seasons must be very slight indeed, even if at all apparent.

While acknowledging the unimportance, in itself considered, of this agency in producing radical changes of climate, it is assumed that, taken in conjunction with the precession of the equinoxes, both combined may have aided materially the dominant cause before mentioned, or alterations in physical geography, in effecting these changes. This motion of precession by which the different seasons of the northern and southern hemispheres are made to coincide successively with all the points through which the earth passes in its path round the sun, combined

* Prin. of Geol., vol. i. p. 286.

with that of the revolution of the apsides, — the latter having the effect to shorten the time in which the seasons complete their revolution, the same being accomplished in 21,000 years, has the result that in 10,500, or one half of 21,000 years, the winter in the northern hemisphere, which now occurs in perihelion, will then occur in aphelion; the points of greatest and least distance of the earth from the sun becoming at that time reversed, and, in another equal interval, returning to the initial point. The argument may be thus stated: When the winter in the northern hemisphere, for instance, occurs in aphelion, at the same time that the earth's orbit exhibits a large degree of eccentricity, the two causes combined will have a material influence in determining a glacial period in that part of the earth, and conversely, whenever the northern summer occurs in perihelion at a time when the terrestrial orbit is exhibiting the minimum of elongation, an interval of unusual warmth in the northern hemisphere must be the result.

In the first place, this hypothesis involves the conclusion already adverted to, that the glacial period in the northern hemisphere was not contemporaneous with that of the southern — a mere assumption which appears to have been suggested for the purpose of avoiding some of the difficulties inseparable from the cosmic winter theory; and secondly, also the absurdity of supposing that the Pliocene Glacial period to which the estimates of geologists accord a lapse of time at least equal to 300,000 years (the actual time being 675,000), can have been consequent wholly or in part on an agency, which, restricted to cycles of 21,000 years, can have aided extreme eccentricity in producing glacial phenomena only for a small portion of the time, being either neutral or antagonistic the remainder. Allowing that the conditions of the Glacial period prevailed for only 300,000 years, over fourteen complete cycles of precession would be accomplished in that time. Eccentricity, as we may see by referring to the table, is also liable to fluctuate nearly from one extreme to the other in periods of 50,000 years; and when we remember that the geological record indicates, with great precision, a uniformly graduated change from the apparent excess of warmth of the later part of the Miocene period to the greatest cold of the Ice era, and from closer proximity to our own times the even more perceptible analogous change from the conditions of that era to those of the present, the hopelessness of the attempt to establish any connection between the Glacial period and these assumed astronomical causes seems sufficiently obvious.

Attaching comparatively little importance to these astronomical agencies, Lyell's chief reliance in the solution of the problem is in the assumed influence which former fluctuation in the physical geography of the globe may have had on its superficial temperature. With candor, admitting that even this cause alone is insufficient to produce the observed effect, and must be taken in connection with the agencies above alluded to, in order to make an approximation to the true theory, he goes on to say that if, after all, "doubts and obscurities still remain, they should be ascribed to our limited acquaintance with the laws of Nature, . . . and should stimulate us to further research," &c.

What a contrast is here presented between this hypothesis, every element of which is involved in doubt and uncertainty, — an hypothesis which depends for support on the fortuitous concurrence of events, that, taken separately, in some instances can have absolutely no effect whatever, and in others only the most inconsiderable, — and the theory in which the single assumption that a well-known movement of the earth is continuous, or persistive, instead of oscillatory, not only solves the problem immediately before us, but by fully harmonizing with the whole arcana of natural phenomena, renders possible a generalization as extensive, at least, and as important as the world of science has ever known.

Deprived of its auxiliary astronomical supports, the theory which seeks to account for former climatic changes by changes in physical geography must be left to stand or fall upon its own merits; and the question, in the form most favorable to the assumption, is, can former vicissitudes of climate have been the result of any conceivable amount of geographical change?

Epitomizing briefly our author's statement of his position, and as far as practicable continuing to adopt his language, we find, under the head of "*Diffusion of Heat over the Globe*," a caution to the effect that, as the climate of Europe cannot be regarded as a type of the temperature which all countries enjoy placed under the same latitude, so the geologist is not hasty to assume that the temperature of the earth, in the present era, is a type of that which most usually obtains, since he contemplates far mightier alterations in the position of land and sea, at different epochs, than those which now cause the climate of Europe to differ from that of other countries in the same parallels of latitude.

"It is now well ascertained," he continues, "that zones of equal warmth, both in the atmosphere and in the waters of the ocean, are neither parallel to the equator nor to each other. It is also known that the *mean* annual temperature may be the same in two places which enjoy very different climates, for the seasons may be nearly uniform, or violently contrasted, so that the lines of equal winter temperature do not coincide with those of equal annual heat or isothermal lines. The deviations of all these lines from the same parallel of latitude are determined by a multitude of circumstances, among the principal of which are the position, direction, and elevation of the continents and islands, the position and depths of the sea, and the direction of winds and currents.

"On comparing the two continents of Europe and America, it is found that places in the same latitude have sometimes a mean difference of temperature, amounting to 11°, or even in a few cases to 17° Fahr.; and some places on the two continents, which have the same mean temperature, differ from 7° to 17° in latitude. . . . The principal cause, says Humboldt, of the greater intensity of cold in corresponding latitudes of North America, as contrasted with Europe, is the connection of America with the polar circle, by a large tract of land, some of which is from three to five thousand feet in height; and, on the other hand, the separation of Europe from the arctic circle by an ocean. The ocean has a tendency to preserve everywhere a mean temperature, which it communicates to the contiguous land, so that it tempers the climate, moderating alike an excess of heat or cold. The elevated land, on the other hand, rising to the colder regions of the atmosphere, becomes a great reservoir of ice and snow, arrests, condenses, and congeals vapor, and communicates its cold to the adjoining country. For this reason, among others, Greenland, forming part of a continent which stretches northward to the 82nd degree of latitude, experiences under the 60th parallel a more rigorous climate than Lapland under the 72nd parallel.

"But if land be situated between the 45th parallel and the equator, it produces, unless it be of great height, exactly the opposite effect; for it then warms the tracts of land or sea that intervene between it and the polar circle. For the surface being in this case exposed to the vertical or steeply sloping rays of the sun, absorbs a large quantity of heat, and raises the temperature of the atmosphere which is in contact with it. For this reason, the western parts of the old continent derive warmth from Africa, which, like an immense furnace, distributes its heat to Arabia, to Turkey in Asia, and to Europe. The north-eastern extremity of Asia, on the contrary, experiences in the same latitude extreme cold; for it has the land of Siberia on the north between the 65th and 70th parallel, while to the south it is separated from the equator by the Pacific Ocean;" * and the author continues at considerable length, commenting upon the above-cited and other like facts, describing the methods by which interchanges of tropical heat and polar cold are affected through the agency of convection, and noting the more remarkable, exceptional instances wherein the climate of certain localities is caused to differ materially from the usual standard by unusual exhibitions and combinations of these causes. The logical conclusion of the whole argument clearly embodies the extraordinary proposition, that the fervid heats of the tropics, and the excessive cold of the poles, are not respectively the results of the direction or degree of inclination, and consequent efficiency of the solar rays, as exhibited in the two localities, but proceed from the irregular action of the convective forces, as modified by the physical geography of the globe. For nothing short of this proposition can afford a sufficient basis for his theory.

The importance of the agencies above mentioned in determining *local* peculiarities of climate cannot be denied.

* Prin. of Geol., vol. i. p. 235–237.

It is equally plain, however, that their effects are of a secondary nature, and can exert a controlling influence only within comparatively narrow limits; for, under the present status, the climate of the several zones must remain essentially the same. The fact that certain exceptional arrangements of the earth's surface are observed to somewhat modify these conditions, does not justify the inference that it may have been possible for the earth's surface to assume such an aspect as to produce a reversal of them, or, at least, a neutralization of their greatest contrasts, in opposition to the primary cause which now determines them. Such a supposition involves the absurdity of a lesser force overcoming and controling a greater. The proposition is self-evident, that while the earth and sun retain their present relative positions, the climate of any given part of the earth's surface must, primarily, be determined by the length of time within the year in which it enjoys the influence of the rays of the sun, and also by the mean angle of inclination those rays make to the plane of its horizon. In other words, as long as tropical areas shall continue to receive the vertical or nearly vertical rays of the sun for the same length of time in the year, and under the same conditions, the sun attained the zenith in each twenty-four hours, no possible changes in the relative proportion of land and sea, or variations in the contour of the land or depth of the sea, one or all, can, in opposition to the solar influence, have the effect to produce there, at ordinary levels, other than a torrid climate. The same elements, on the other hand, conversely employed, determine the temperature at the poles. As long as the present status continues, no effect of any possible rearrangement of the physical aspect of the earth's surface upon the ordinary heat-distributing agencies can avail to reverse, or even modify generally, the prevailing climatic conditions of the present time.

Equatorial and polar areas present the extremes of terrestrial climate, and the well-known property of heat, by which it is continually seeking an equilibrium, must, of course, superinduce a tendency towards a mutual interchange of heat and cold between them. This tendency is effected, as far as is possible under the circumstances, by means of aerial and marine currents, the heat-distributing agents above alluded to. The aerial currents, or winds, not only convey large amounts of sensible heat, radiated from the heated earth-surfaces of the tropics towards the poles, but they are also largely charged with water in the gaseous form, containing heat in its latent state, which becomes sensible as the vapors encounter the colder winds of high latitudes. There is thus a constant tendency towards an equalization of heat and cold between the two extremes; and were they in close proximity, the heat of the tropics and the cold of the frigid zones would largely neutralize each other. But the point, longitudinally, of greatest heat — the equator — is separated from the points of greatest cold — the poles — by distances of about 6,000 miles, nearly one half of which intervenes between the tropical and polar circles. These distances are far too great to admit of the transference of heat by the winds (which readily part with any excess), from one point to the other, in quantities sufficiently large to produce sensible effect; and ocean currents, although retaining heat with greater tenacity, are strictly localized, affecting sensibly only very limited areas of the extreme zones. Even the temperate zone, which, from its position, may be termed the battle-ground between the contending forces of equatorial heat and polar cold, although, from the fact of its constituting such battle-ground, exhibiting marked diversity of climatic conditions, according as one or the other of those adverse influences may predominate, bears, on the whole, the general climatic characteristics determined by the duration and intensity of the solar influence. Each zone receives annually from the sun a determinate quantity of heat, and that quantity determines, approximately, the climate of such zone; and the amount of modification effected by geographical causes, and the *irregular* action of heat-distributing agents, must be, especially in the extreme zones, extremely small. It should be observed, perhaps, that the *regular* action of those agents tending to equalize the temperature of the whole earth, whatever the amount of its effect, is not necessarily to be taken into account, that being a stable element in determining the actual difference in mean temperature between the extreme climates.

Lyell affirms the antiquity of the existing continents.

From the commencement of Tertiary times, the geographical aspect of the surface of the earth has at no one period differed to any great extent from that of another. Oscillations of level, confined to narrow limits, have occurred, land giving place to sea, and sea to land; but the distinct sets of plants and animals by which the principal masses of land are inhabited, show that they must have been separated from each other for an immensely long period of time. Throughout the whole extent of Tertiary times, probably, and certainly within so much of the same as will, at least, include the Pliocene Glacial period, there have been no geographical changes in the northern hemisphere of sufficient importance to disturb in the least degree the conditions on which depends the marked contrast between the climates of Europe and North America; a difference none will probably deny to be mainly due, as our author asserts, to the peculiar geographical aspect of those portions of the earth. The fauna and flora of times anterior, as well as the monuments of the Ice period itself, which exhibit all the "signs of glaciation, such as erratic blocks, scored surfaces of rock, striated boulders, and deposits filled with arctic species of marine shells, which are to be seen in full force on the North American continent ten or more degrees farther south than in Europe," * show that these geographical conditions have remained essentially the same, certainly throughout the whole Glacial period, and most probably during the immense lapse of time comprehended between the close of the Secondary Age of the earth and the present time.

The traces that now remain of the Glacial period point to a time the climatic conditions of which differed from those of the present, we may almost say, in the greatest possible degree. Now, according to our method, there have been at least three of these ice periods, or one to each geological epoch, within Tertiary times. We are absolutely sure there has been one such period within that space of time. If, then, there has been no appreciable geographical change during a period marked by the most pronounced climatal change the geological record bears evidence of, the utter futility of the attempt to make the latter dependent upon the former becomes obviously manifest.

* Prin. of Geol., vol. i. p. 288.

The extraordinary manner in which the isothermal lines of polar and upper temperate areas diverge from parallelism with the lines of latitude, arises undoubtedly from the modified action of the heat-distributing agencies before mentioned, such modification being more or less due to geographical causes. As we approach the equator, however, these divergences rapidly diminish, and, on reaching the tropics, the isothermal lines become, comparatively speaking, nearly parallel with those of latitude. Such being the case, were we even to admit that the semi-tropical climates which circumpolar countries enjoyed in former times were the result of fluctuations in physical geography, such hypothesis will not suffice for the quite as pronounced climatal changes that have occurred within the tropics. At the present time, the terrestrial surface is quite as diversified in low as in high latitudes, and the heat-distributing agents are quite as active there; and, according to Lyell's hypothesis, the former should exhibit relatively the same local climatal differences as the former. Such not being now the case, if we assume all the other conditions determining climate to have been the same in former times as now, the climate must have been the same; for like causes must in the nature of things always produce like results. From the observations of Agassiz, we know that the climate of tropical countries has, in the past, differed essentially from that of the present; and it follows with absolute certainty that there must have been modifications of the primary or controlling cause determining climate. The effect of geographical causes, which in high latitudes exert but a secondary and local influence, becomes, in the tropics, almost completely neutralized under the potent influence of the dominant primary cause — the heat of the sun. If, then, the peculiar phenomena of the Ice period invaded the tropics, and in all probability reached the equator, as indicated by the observations of Professor Agassiz, such a radical divergence from the present climatal conditions of the torrid zone cannot have proceeded from a cause whose effects are now almost entirely confined to high latitudes, and even there are only competent to effectuate comparatively trivial local modifications of the conditions primarily determined by the dominant controlling solar influence.

It is indeed surprising that an extended argument is necessary to establish the truth of the proposition that the climate of any given latitude of the earth's surface is directly determined by the amount of light and heat it receives annually from the sun, and the mode in which such amount is distributed over the several seasons of the year — a proposition which, as we fix our attention upon it, cannot but assume an axiomatic aspect. A practical demonstration of its truth lies in the fact, that there is an invariable, fixed rate at which the terrestrial temperature is observed to diminish as we approach the poles from the direction of the equator, whatever may be the amount of exceptional local deviations from the mean, due to causes of a secondary nature. The truth of the above proposition being admitted, the radical changes in the climate of the earth that have occurred in former times, necessarily involve changes in the mode under which the solar influence is distributed over its surface. If a movement of the earth can be shown to be in progress, the result of which would be to effect precisely such changes in the method of distribution as shall reasonably account for former climatal changes, nothing more seems required upon which to base a general theory of climate.

If, on the other hand, we refuse to admit these elements, and endeavor to make geographical change the prime factor in the solution of the problem, we encounter the task — one which cannot be avoided — of showing how the last-mentioned agency, in opposition to the solar influence, can so operate as to neutralize, or even modify, under the present status, the general and uniform decrease of temperature now observed to prevail from the equator to the poles. It will not suffice to demonstrate that a certain arrangement of oceans and seas, continents and islands, would have the effect to modify the climatal conditions which must ensue in the absence of such arrangement. A theoretical distribution of sea and land must be made, such as can be shown, from observed effects of the present arrangement, adequate to produce, at least, a neutralization of the contrast between equatorial and polar climate. A geographical combination is demanded, not that shall merely tend to equalize the temperature of limited areas, separated by a few degrees of latitude only, but which, by an equalization of mean temperature, shall establish a *climatal analogy* throughout all latitudes. It can scarcely be necessary to observe that we can rationally conceive of no such combination capable of producing, under existing conditions, even a modification of the mean rate at which the terrestrial temperature decreases from the equator to the poles; and, as repeated instances have occurred in the past history of the earth, in which the temperature of at least the whole northern hemisphere has approximated a mean, difference of latitude making itself felt, not by difference in the annual mean of heat, but by its greater or less concentration within a portion of the year, it follows that geographical change cannot possibly be assumed as the dominant cause of this class of former climatal changes. Being thus utterly inadequate to produce the repeatedly recurring instances of equalization of the mean annual temperature throughout all latitudes, it becomes impossible to accord to it more than a subordinate rank among the agencies affecting the terrestrial climate.

CHAPTER VI.

ON THE PROPER AND DETERMINATE CAUSE OF PRIMARY CLIMATAL CHANGES.

THE SUN. — SPECULATIONS CONCERNING THE AGE OF THE EARTH. — DIMINUTION OF THE OBLIQUITY OF THE ECLIPTIC. — TRUE CAUSE OF DIMINUTION. — EXAMINATION OF THE HYPOTHESIS WHICH CONNECTS THE STABILITY OF NATURE WITH THE CONTINUANCE OF THE PRESENT INCLINATION OF THE EARTH'S AXIS. — THEOREM OF LAGRANGE. — PARALLEL BETWEEN THE "STABILITY OF NATURE" THEORY AND THE PTOLEMAIC ASTRONOMICAL SYSTEM.

If it has thus far been made to appear that the subordinate and local agencies, heretofore considered, are entirely inadequate to produce the climatal phenomena indicated in the geological record, there appears to be no rational alternative but to consider the radical climatal changes characterizing former eras in the earth's history as the effects of variation in the method by which the distribution of the solar influence over the terrestrial surface has been effected. In this connection the nature and constitution of the sun becomes an interesting subject of inquiry.

This vast body, constituting the central orb of the solar system, — whose diameter may be stated at 883,000 miles, whose volume exceeds that of our earth nearly a million and a half times, and that of all the planets together seven hundred times, — which, by its prodigious bulk, controls, through the law of gravitation, the motions of all the other bodies of the solar system, retaining them, each in its proper orbit, is also the source from whence emanate those inexhaustible stores of light, heat, and magnetism, by which the earth is constituted the abode of life. To the eye of man, presenting the most conspicuous and glorious object in nature, the most unremitting efforts have been made to acquire some definite knowledge of its constitution; but, up to the present time, these efforts have met with but indifferent success. The immense distance which separates us from the sun, were all other circumstances favorable to observation, would preclude the idea of an intimate acquaintance with its superficial aspect; for when we consider how little, comparatively, can be learned of that planetary body nearest the earth, — the moon, — with the assistance of all the improved appliances of modern times, the attempt might well be abandoned of investigating, by visual observation, the physical properties of an orb nearly four hundred times further removed from us. A still greater obstacle to solar observation is the fiery appearance of the sun's disk, the same requiring the interposition of substances, more or less opaque, between it and the eye of the observer, which tend to obscure the vision in the same degree that they intercept the excess of light. Excepting the appearance of spots upon the sun's disk, supposed to be breaks or openings in the luminous envelope of phosphorescent clouds by which the supposed solid nucleus is surrounded, and elevations and depressions in those clouds, but little can be learned of the sun by actual observation. No theory of the sun that has as yet been advanced is free from serious objection. For our present purpose, however, speculations as to how such a prodigious emission of heat is maintained are of less moment than are inquiries concerning the stability and permanence of the calorific effect of the solar influence at the surface of the earth. How long has the sun diffused light and heat throughout the solar system, and over the earth? Is the amount of solar heat a fixed quantity, or is it subject, at times, to variation in sufficient measure to affect the well-being of the inhabitants of the planets?

As it is impossible to compute the distance from the earth of nearly all the innumerable host of fixed stars that stud the heavens, they being situated beyond the utmost limit of human calculation, so also there is a bound beyond which the geologist can hardly hope to pass in his investigations concerning the age of the world. While we can hardly doubt that the sun has shone upon the earth from the time of its creation, or, rather, from the time it assumed its present form and position in the solar system, the first positive evidence we have of the actual presence of the solar influence upon the earth is found in the inception of organic life upon its surface, and the initiation of the peculiar order of arrangement that surface has assumed, through meteorological process, consequent on changes of temperature. In other words, it may be conceived possible to approximate the age of the world from the commencement of the earliest geological period that can be traced in the earth's crust; the influence of the sun being further indicated wherever traces of life exist. Beyond this point in the history of the earth, all must be conjecture. We may imagine it, for instance, an erratic body, perhaps an exhausted sun, wandering through space, a true planet, coming at last within the sphere of attraction of the immense mass that now controls its movements, or we may suppose it to have been a fiery spark of the material of the sun itself, struck off and projected from that body, a mass of incandescent matter, requiring ages of inconceivable duration in which to complete the cooling-off process to the degree necessary to form a solid, superficial crust adapted to receive the germs of life.

Aside, perhaps, from the question of origin, a considerable degree of probability, it has been supposed, attaches to the latter hypothesis. If a heated, incandescent, original condition can be assigned to both the earth and its satellite, the present superficial aspect of the moon may, from the entire absence, as far as we know, of all abrading and disintegrating forces at its surface, be almost if not precisely the same as it exhibited on first assuming a solid form. The surface of the earth, at the same stage, would present an analogous appearance. Lofty mountains and deep depressions of solid plutonic rock — solid save the rents and fractures caused by the contractions attendant on the cooling process, aggravated perhaps by violent volcanic forces — must have marked the incipient formation of the earth's crust, as well as that of the moon, although, perhaps, in a more excessive degree in the former on account of the greater volume of the heated mass. The superior bulk of the earth, largely in excess of that of the moon, may be supposed to have attracted to itself and deprived the latter of those aqueous and aerial elements and compounds, which, during the period of incandescence, must have existed in a highly-attenuated, gaseous form, absorbing thus to its own use those elements of superficial change, and dooming the satellite, after the expenditure of its primitive volcanic or other forces, to a condition of perpetual superficial repose. If volcanoes and earthquakes have not become obsolete in the lunar economy, they are the only remaining means whereby changes in the aspect of the moon's surface can be effected; but even these agents, if now in operation, can only rend and contort the rocky mass: they cannot disintegrate, pulverize, and reduce it to soil; and, in the absence of more potent forces, the surface of the moon must continue to exhibit, while remaining in its present relations, essentially the appearance it did on first assuming the solid form.

But the chaotic desert of rock which constituted the primitive surface of the earth, as soon as it became sufficiently cooled to admit of the condensation and precipitation of the aqueous and aerial gases, became the object of attack, in addition to those above named, of a great number of destructive forces, of both a chemical and mechanical nature. The incessant action of these influences, proceeding with more or less rapidity, according to circumstances, breaking up and pulverizing, disintegrating and dissolving, the solid rocks, and spreading the comminuted material over the original uneven and rugged surface, is the agency which brought it, in course of time, into a suitable condition to receive the germs of life, and eventually into that state of teeming fertility and beauty we now behold.

However probable the conjecture that the sun was diffusing light and heat during the whole of this time, there is no absolute certainty of it. But, as its influence was equally indispensable to organic life, and as the primary agency in inducing superficial change, we may be sure that it has shone from the time of the first appearance of these phenomena upon the earth. The earth's crust, according to Beaumont and other geologists, comprises thirteen distinct revolutions, formations, or periods, — whichever term we choose to employ, — or, rather, twelve, if we are justified in assuming Post-Tertiary times as a portion of the Pliocene epoch. There are, however, acknowledged breaks in the order of succession of these revolutions, which represent, undoubtedly, unknown periods or revolutions, the traces of which have been entirely erased from the geological record; and there can be no doubt but thirteen is the least number that has actually taken place. The question now recurs, What is the amount of time represented by one of these geological periods?

Geologists universally decline to hazard a conjecture, although they are agreed it must be immensely great. In order to reach an approximation, we are compelled to resort to data furnished by the new system. A geological period must be the result of a half revolution of the earth, transverse to the axis of its diurnal rotation, or whatever portion of such revolution will include a complete glacial period and its complementary interval of repose. On the supposition that the value of the transverse rotary motion of the earth is constant at 48″ per century, its present rate, an interval of 1,375,000 years is the time required to accomplish a single geological revolution or period; the aggregate for the thirteen being 17,875,000 years. In view of possible retardations of the transverse rotary motion of the earth, and also of the probability that the centennial rate is somewhat over-estimated, the above figures are to be regarded as a minimum; the actual time which has elapsed since the deposition of the earliest known fossil-bearing strata evidently being considerably in excess of that amount. Adopting it, however, as the basis of our calculation, we may suppose that another equal period of time was no more than sufficient for the development of the primordial germs of organic life into the forms first preserved and transmitted to us. For thirty-six millions of years, and, more probably, fifty millions, the effect of the light and heat of the sun at the surface of the earth must have approximated its present value, such approximation being absolutely essential to organic existence.

That these calculations, in common with all others of a similar nature, are of a somewhat hypothetical character, must be freely admitted; while their moderation, as contrasted with those others — with which, it may be observed, they are not inconsistent — is evident. "Sir William Thomson," according to Professor Huxley, "believes that he is able to prove, by physical reasonings, 'that the existing state of things on the earth, life on the earth — all geological history showing continuity of life — must be limited within some such period of time as one hundred million years.'"* The time, therefore, during which the solar influence has been enjoyed at the surface of the earth, in due proportion for the development and support of life, may be estimated as somewhere from thirty-six to one hundred million years.

If, in all this inconceivably vast interval of time, no variation in the absolute amount of light and heat received by the earth from the sun can be detected, of sufficient importance to affect the continuity of life, how can we avoid the conclusion that, in the attribute of stability, resembling its great Designer and Creator, the sun, from the moment of its creation down to the present time, has remained unchanging and unchanged, and in the future will so continue, until the great Arbiter of the Universe shall decree its final dissolution? Assuming the evidence adduced as bearing upon the subject to be inconclusive, or, even in the entire absence of any such evidence whatever, how natural it is for the human mind to accept this, to our eyes, the most prominent and glorious object in the material universe as the physical emblem of God, and to believe that He, during its existence, has endowed it with his own attributes of unchangeableness and beneficence. The idea of vacillation and change, as connected with the sun, is quite as repugnant to our reason as would be the ascription of the same to the Creator; and this innate intuition, in addition to the proofs we are able to advance, leads us with the utmost confidence to the conclusion that the sun, as a source of light and heat, is an unchangeable body.

Nor is this conclusion affected by the fact, that, to the present time, notwithstanding the new light afforded by spectroscopic analyses, science has been unable to render any satisfactory account of the origin of the solar light and heat; nor, further, by whatever questions may arise respecting the probable future duration of our great luminary. Whether, as some have supposed, it contains in its colossal reservoir a sufficient store of light and heat to furnish the solar system with an undiminished supply for millions of centuries, or not, however interesting as sub-

* Lay Sermons, Addresses, and Reviews, p. 244.

jects of scientific inquiry, are questions that have no practical bearing on the issue involved in this discussion. We may be permitted, however, to observe on this point, that, according to Guillemin, even if "the solar globe loses its heat year by year, it is perfectly certain that there will still sufficient heat be left to support life on the earth and the other planets for millions of years to come." If the sun is experiencing, as above suggested, a gradual decrease in temperature, the fact that for forty, fifty, or perhaps a hundred million years it has maintained a degree compatible with life upon the surface of the earth, shows that the amount of decrease, if any does actually exist, must be extremely small; so small, indeed, that it need not be taken into account as affecting the truth of the proposition that the sun, as a source of light and heat to our planet, is an invariable body.

The solar influence being thus invariable, — the mean distance between the earth and sun remaining continually the same, — and the earth, in consequence of its spherical shape, presenting always the same extent of its surface to the sun, it would seem to follow that the mean annual amount of heat received by the earth from the sun would be a constant quantity. Geometers, however, as stated in the preceding chapter, have somewhat qualified this conclusion, making the mean of heat received to vary in a ratio corresponding to certain changes that occur in the minor axis of the terrestrial orbit. The amount of thermal change involved is so small, — being but about three one-thousandths of the whole sum, — that it cannot appreciably affect our results; and we may assume that the annual mean of solar heat received upon the surface of the earth is also invariable.

Were the earth to remain stationary with regard to the sun, neither revolving about the same, nor turning on its own axis, there could be but little variation in the temperature of the different portions of its surface. One hemisphere — that turned away from the sun — would experience an eternal night and an eternal winter, while the other would enjoy unending day and summer during the continuance of such relations. The centre of the illuminated disk would receive by far the largest share of the solar influence, the amount gradually decreasing to the circumference, where it would scarcely exceed that of the opposed side. Let, now, the diurnal rotation be supposed to commence, and the centre of the disk, or point of greatest heat, would traverse the circumference of the earth, describing the line of the equator, and effecting an equal daily distribution of the solar influence throughout its whole extent. The same effect, relatively proportional to the quantity of heat received, would follow at points situated at a distance from the equator, on either side of it, upon lines parallel to the same, or lines of latitude. One result of the motion would be to determine a division of the time of revolution into two equal portions, constituting the day and night. The amount of the solar influence received at any given point on the earth's surface would be a fixed quantity, determined by the distance of that point from the equator. These conditions must, of necessity, continue to prevail as long as the status should remain unchanged.

If we now imagine the generation of a motion of the earth around the sun, and that the plane of the path traversed in the revolution be perpendicular to the axis of diurnal rotation, no disturbance will arise in the abovementioned terrestrial conditions in consequence of such motion of revolution. The time of revolution, or year, would exhibit no succession of seasons such as we now experience, and the climate of each and every part of the earth's surface, as determined by latitude, would remain perpetually the same; the tropics experiencing perpetual summer, the poles perpetual winter, and intermediate latitudes every stage of gradation between the two.

Let us now further suppose the axis of diurnal rotation to become inclined to the orbital plane, the only possible method under the present arrangement of the solar system whereby a variety of seasons, as now seen, could be effected. At the angle of inclination which now prevails, — about 23° 28′, — the result would be the diurnal and annual variations of the present time, *and no other*. None other could be possible. Any other given degree of inclination would produce another and different stated condition of climate and seasons, and whatever the degree of inclination, the resulting conditions must, in the nature of things, remain unchanged so long as the producing cause or inclination should continue. Each and every position it may be possible for the terrestrial axis to assume, by varying the manner in which the annual supply of solar heat should be distributed over the earth, would determine climatal conditions peculiar to such position. It follows, therefore, with the utmost certainty, that if the climatal conditions of former eras have differed in any observable degree from those which now prevail, or from each other, such differences can have been due only to the cause above indicated. We can hardly conceive of greater changes or more marked climatal contrasts than those which would be presented, first, by a change from the present climate of arctic localities to one adapted to sustain there an abundant and luxurious vegetable growth; secondly, from that now seen in the tropics to another capable of originating there a system of glaciation in mountains of moderate elevation; or, more remarkable than either, perhaps, a variation from the present climate of the whole earth, which exhibits a uniformly graduated decrease of temperature from equatorial heat to polar cold, to another, in which this relation should become neutralized, so that a close approximation to *climatal analogy* should prevail throughout all latitudes. The geological record indicates in no ambiguous or uncertain manner, that all these, and other radically distinct climatic states, have prevailed over and over again in their respective localities; and these states can have resulted only from different positions of the earth's axis, entailing changes in modes of distributing the annual amount of solar heat over the earth's surface. Sudden and violent changes of inclination, by inducing corresponding variations in climate, must have involved wide-spread destruction among the living inhabitants of the globe, of which there is no evidence, and they are, moreover, contrary to the analogies of nature. It would appear, then, that as such climatic changes have occurred in the past, involving variation in the position of the terrestrial axis, such variation must proceed in a uniform, constant, and extremely slow manner. Is there any motion of the earth's axis, or rather of the earth, to which we can refer these vicissitudes of climate?

Diminution of the Obliquity of the Ecliptic. — To the question above propounded, the observations of astronomers, made at frequent intervals throughout a period of about three thousand years, have furnished a positive, unequivocal answer in the affirmative. These observations show that the earth is now, and has been for at least thirty centuries, subject to an exceedingly slow but continuously progressive rotatory motion, at a right angle to the diurnal revolution. One effect of this motion being to vary the relative positions of the equatorial plane and the plane of the terrestrial orbit, astronomers have been accustomed to allude to it as "the diminution of the obliquity of the ecliptic." In his "Outlines of Astronomy," section 640, Herschel thus alludes to this motion: "Meanwhile there is no doubt that the plane of the ecliptic does actually vary by the action of the planets. The amount of this variation is about 48″ per century, and has long been recognized by astronomers by an increase of the latitude of all the stars in certain situations, and their diminution in opposite regious. Its effect is to bring the ecliptic by so much per annum nearer to coincidence with the equator." According to Dr. Brewster, as cited by Vose, "the obliquity of the ecliptic to the equator, was long considered a constant quantity. Even so late as the end of the 17th century, the difference between the obliquity, as determined by ancient and modern astronomers, was generally attributed to inaccuracy of observation, and a want of knowledge of the parallaxes and refraction of the heavenly bodies. It appears, however, from the most accurate modern observations, at great intervals, that the obliquity of the ecliptic is diminishing." * In order to give the reader a general idea of diminution, the table on page 198, from Hinds' "Solar System," is inserted.

It can hardly be supposed that the more ancient of these observations were made with the degree of accuracy characterizing those of modern times. The Arabian astronomers seem to have ignored quantities of less value than minutes, and the figures set against the names of some who have been regarded as the highest astronomical authority of their times, cannot but suggest an appropriation of the labors of preceding observers. But, notwithstanding, the regular progress of the movement is apparent. The difference between the first and largest inclination,

* Vose's System of Astronomy, p. 144.

Table exhibiting the Principal Determinations of the Obliquity of the Ecliptic in Ancient and Modern Times.

B. C.	OBSERVERS.	°	′	″
1100	Tcheou-Kong, ...	23	54	2
324	Pytheas of Marseilles, ...	23	49	20
230	Eratosthenes of Cyrene, ...	23	51	15
140	Hipparchus.	23	51	15
50	Lieou-Hang, ...	23	45	39
A. D.				
140	Ptolemy, ...	23	51	15
173	Chinese observations, ...	23	41	33
461	Tsou-Chong at Nanking, ...	23	38	52
629	Litchun-Fong, ...	23	40	4
830	Almanum, ...	23	33	52
879	Albategnius at Aracte, ...	23	35	0
987	Aboul Wefa at Bagdad	23	35	0
995	Abul Rihau, ...	23	35	0
1080	Arzachel in Spain, ...	23	34	0
1279	Cocheu-Kong, ...	23	32	12
1303	Prophatius, ...	23	32	0
1430	Ulugh Beigh at Samarcand, ...	23	31	48
1460	Regio Montanus (Tables), ...	23	30	0
1587	Tycho Brahe, ...	23	31	30
1660	Hevelius, ...	23	29	30
1690	Flamsteed, ...	23	28	56
1750	Bradley, La Caille, &c., ...	23	28	19
1769	Maskelyne, ...	23	28	10
1800	Delambre and others, ...	23	27	57
1825	Bessel, ...	23	27	43.4
1840	By observation at various places, . .	23	27	36.5

viz., 23° 54′ 2″, the result of the observations of Tcheou-Kong, 1100 years B. C., and the smallest, 23° 27′ 36.5″ as obtained in the year 1840, is 26′ 25.5″; which would indicate a mean centennial decrease of about 52.8″.

Dr. Brewster, by comparing about 160 observations made by both ancient and modern observers, obtained as a result 51″ as the diminution per century. Later authorities, however, from computations based upon the most exact modern observations, have determined 48″ to be a still nearer approximation to the true value, which number, although liable to still further reduction, has been adopted as the basis of our calculations in estimating the duration of the various epochs in the past history of the world.

We know, therefore, — and our knowledge is of the most positive character, — that within the past thirty centuries the poles of the earth have turned or moved in a direction at right angles, or transverse to the axis of the diurnal revolution, to the amount of about two-fifths of a degree; a movement which, if assumed to be continuous, must determine, in stated times, complete revolutions of the earth. In addition to our positive knowledge on this point, satisfactory evidence exists that the motion alluded to was known to be in progress by ancient astronomers long before the date of the earliest written records that have been transmitted to our times.

Allusion has been made in a former chapter to a tradition, or theory, current among ancient Eastern nations, to the effect that the ecliptic was once perpendicular to the equator, dividing the year, at the North Pole, into a long day and night. It would be most absurd to suppose they could have attained to that belief by other means than an acquaintance with the astronomical fact of diminishing obliquity; and they must, undoubtedly, have arrived at a knowledge of that fact through the use of precisely the same means that more modern nations have employed. Observations must have been multiplied, differences in results noted and transmitted from one generation to another, resulting at length in the discovery of the fact of motion, and, at last, in the development of the theory embodied in the tradition — a natural deduction from that fact. There is certainly no reason to suppose that they could arrive at a knowledge of an astronomical motion of this nature within a less interval of time than the moderns. If it has taken the latter three thousand years to become acquainted with the fact that obliquity is diminishing, it must have taken the ancients at least an equal length of time; and we may fairly assume that mankind have been multiplying observations, and collecting data in relation to obliquity, for at least six thousand years, and that such observations and data have all tended to show a continuous diminution in the same, involving an amount of rotatory motion, on the part of the earth, equal to four-fifths of a degree.

In our geological inquisition, we found the crust of the earth teeming with phenomena utterly inexplicable under any hypothesis save that of the transverse rotation of the earth; and, on the other hand, not a single fact, which, carefully considered, assumed even the appearance of inconsistency therewith. All the effects of such rotation that we could reasonably expect to encounter are to be found, not in single isolated instances, but in analogous groups, following each other within those groups in regular order and sequence.

Now, when, aside from these incontestable geological evidences of this motion of the earth, we arrive at a knowledge of it through another independent and totally distinct process, — when we discover by astronomical observation that such motion has been in progress certainly for three, and probably for six thousand years, and that it has been tending continually one way, — what more is required to establish the transverse rotation as an essential part of the terrestrial economy? Nothing is wanting; and we may as confidently assert that the earth describes complete revolutions transverse to the diurnal rotation in some such periods of time as two and one half or three million of years, as that it revolves on its polar axis at intervals of twenty-four hours, or about the sun in those of one year.

Nor will the truth of this conclusion be in the least affected by either success or failure in the effort to assign a proper cause for the motion. A simple and apparently a just method of accounting for it is to be found ready at hand. Authority, however, has rejected the element on which the solution depends; and if, in examining the grounds of this rejection, it is found to rest upon a substantial basis, nothing remains but to search for some other rational mode of solution.

Lunar Attraction on the Accumulation of Matter at the Earth's Equator, the Cause of diminishing Obliquity. — It is a well-ascertained fact that the equatorial exceeds by twenty-six miles the polar diameter of the earth. The centripetal force of the diurnal revolution tending to withdraw material from the poles and heap it up at the equator, has accumulated there a belt or girdle, estimated to be 25,000 miles long, 6,000 miles wide, and 13 miles deep. In order to bring the earth in shape to a perfect sphere, 1,950,000,000 cubic miles of material would have to be removed from equatorial latitudes, a quantity which would form a globe 1550 miles in diameter, — that of the moon being about 2150, — the proportional volume of which globe to that of the moon being as about 1 to

2.8, and as compared with that of the earth, about one 137th of its whole mass. Even this amount, vast as it is, is not all; for according to Lyell, it has been inferred from certain astronomical data, "that the equatorial protuberance is continued inwards; that is to say, that layers of equal density are arranged elliptically, and symmetrically, from the exterior to the centre." *

The varying force of the moon's attraction, in different parts of its orbit, upon this belt of material, is assumed to be the cause of diminishing obliquity.

In a treatise on astronomy by John Vose, published at Concord, N. H., in the year 1827, is to be found, on page 145, the following solution of the problem: —

"The attraction of the moon on the spheroidal figure of the earth, affords so natural an explanation of the cause of diminution, in the obliquity of the ecliptic, that it is wonderful any other should have been sought.

"Let T (of the following diagram) be the earth, M the moon, N S the earth's axis, E Q the equator; the line T M

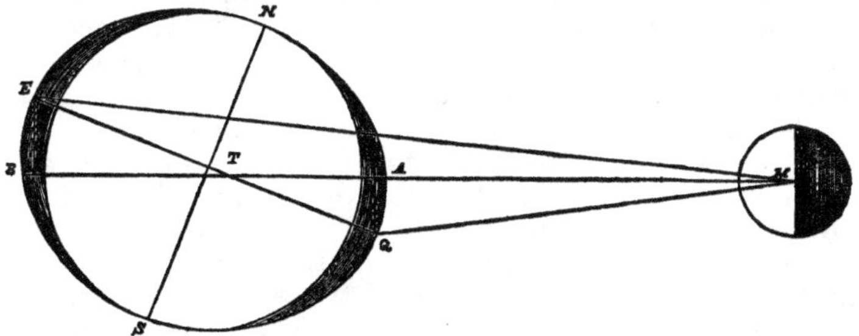

a radius of the moon's orbit at a node, or when it coincides with the plane of the ecliptic; A B the diameter of the earth, as cut by the plane of the ecliptic. In the triangle A M Q, the line M Q may represent the force of the moon's attraction on the accumulated matter of the earth, at the equator, on the side next to the moon. This force, by the principles of motion, may be resolved into two other forces,† represented by the lines A M and A Q; the former of which, being in the plane of the ecliptic, cannot affect

* Prin. of Geol., vol. ii. p. 202.

† "Enfield, Mechanics, Book ii., Ch. iii., Prop. xvi."

the inclination; but the latter operates to diminish the obliquity. This force must act in every part of the moon's orbit, except at the beginning of Aries and Libra.

"The action of the moon on the opposite side of the earth, must be counter to that we have considered. But, from the well-known principle, that the force of gravity diminishes as the squares of the distances increase, the effect on different sides of the earth must be unequal, and least on that side, which is opposite to the moon. But if the force of the moon's attraction on the different sides of the earth were equal, the counteraction on the opposite side must be less than the diminishing action on the side of the earth next to the moon; for the line B E is equal to A Q; but the line B M is longer than A M. If therefore B E and B M represent a force equal to A Q and A M, as in the hypothesis, B E must be less in proportion to the whole than A Q; B E being less in proportion to B M than A Q is to A M. Unequals being taken from equals, the remainders are unequal.

"The inclination of the moon's orbit to the plane of the ecliptic must cause her action to be greater at some times than at others; but cannot prevent her operating in every revolution to diminish the obliquity.

"The attraction of the sun on the matter accumulated at the earth's equator must produce an effect similar in kind to that of the moon. But the distance of the sun from the earth is so great, that the line A Q bears a very small proportion to the line Q M or A M. The attraction of the sun also, in different parts of the earth, becomes almost equal, as in the case of the tides. The effect of the other planets on the obliquity, must be extremely small.

"If the explanation here given of the cause of the diminution in the obliquity be just, it can neither become stationary nor increase without power extrinsic to the solar system; but must continually decrease, and in time become extinct;" and thus, having reached the point at which the motion under consideration has brought the earth's axis perpendicular to the ecliptic, our author pauses, evidently somewhat bewildered at having penetrated so far into the future. There is no reason, however, that the movement should become stationary, or retrograde, at this point more than at any other; for, if the cause continues constantly in operation, the effect must continue to follow indefinitely, and no sooner would obliquity cease on one side of the equator, than it would recommence upon the other.

It appears to me very absurd to attribute this motion of the earth to any other agency than that above indicated. The proportional attractive influence of the various masses of the solar system upon the matter of the earth is clearly indicated in the case of the tides. The moon, notwithstanding its inferiority of bulk, from its close proximity to the earth, is found to exert the controlling influence in determining that movement of the waters of the ocean; while the sun, although almost infinitely exceeding the moon in volume, from being situated nearly four hundred times further off, exerts but a secondary influence, serving only to retard or accelerate, in a comparatively feeble degree, the primal impulse of the lunar attraction. Unquestionably, the sun attracts the earth, *as a whole*, more powerfully than the moon does; but, as we see in the case of the tides, it has a much less influence in determining *local* perturbation. The influence of the planets upon the tide is so small as not to be taken into account.

Now, common sense would appear to dictate, that the same proportional degrees of attractive influence which the masses of the solar system exert on the tides, would be exhibited in the strictly analogous case of the equatorial protuberance. But Herschel, as before cited, asserts that the variation of obliquity is due to "the action of the planets;" and as this variation and perturbation in the eccentricity of the Earth's orbit are ascribed to a common cause, and as the latter are held consequent on the attraction of the nearer and larger planets, Jupiter and Saturn playing the principal part, and Venus and Mars also exerting a sensible influence, we may infer that it is the influence of these four planets to which he ascribes the diminution of obliquity.

From the considerations above suggested, the opinion seems more probable that the conjoint attraction of these planets, while sufficiently powerful, when acting upon the terrestrial mass *as a whole*, to produce sensible changes in the form of its orbit, is obviously as inadequate to exert an appreciable influence upon a mere inequality of its surface, such as the equatorial protuberance, as it is to affect the tides; in other words, these planets can perceptibly attract the whole earth; but when a fraction of its mass, amounting to about one one-hundred and thirty-seventh part, is so disposed about its equatorial circumference, that a portion of such fraction is subject to their attractive influence under circumstances slightly different from another portion of the same, no appreciable effect is possible.

This point may be made clear by comparing the quantity of matter composing these planets, and their distances from the Earth, with the quantity of matter and distance of the Sun from the Earth.

We find by referring to Astronomical Tables, that assuming the quantity of matter composing the Earth as unity, the proportional quantity of the Sun is 333,928; and its distance from the Earth 95,000,000 miles. In the same ratio, the proportional quantity of matter in Venus is 0.8899; in Mars, 0.0875; in Jupiter, 312.1; and in Saturn, 97.76; the aggregate being 410.8374, or less than one eight hundred and twelfth part that of the Sun. If, therefore, these four planets constituted but one body, and that body remained stationary with regard to the Earth, at the same distance from it that the Sun is, such body would exert less than $\frac{1}{812}$th part of the attractive force of the Sun upon it. But Jupiter, which contains over $\frac{3}{4}$ of the material of the four planets, never approaches the Earth nearer than about four times the distance between the Sun and Earth; and Saturn, constituting nearly $\frac{99}{100}$ of the remainder, is nearly twice as far removed from it as Jupiter. Of the two other planets, so insignificant in magnitude, comparatively, the orbit of one is about 27,000,000, and that of the other about 49,000,000 of miles from that of the Earth. And, besides, the occasions must be rare, indeed, when these planets can exert a conjoint influence upon the Earth by occupying points in their orbits on a line with the Earth and Sun, and on the same side of the Sun; and lastly, it is physically impossible that the four planets named can attract the Earth in conjunction upon the same side of the Sun; for the action of Venus, which, excepting the Moon, is the nearest to the Earth of all the planets, and exerts, consequently, according to magnitude, a proportionally greater attractive influence upon it, being that of an inferior planet, must be counter to the attraction of the larger and more remote superior masses.

It is, indeed, conceivable that all of these four planets may assume positions in their orbits, so as to exert a conjoint attractive influence upon the earth; but such a contingency, although in the range of possibility, may, at the same time, never have transpired since the solar system first assumed its present order of arrangement. But even if it were of frequent occurrence, as it could only take place when the superior planets were, approximately, at their greatest distances from the earth, — an increase of some 200,000,000 miles over the minimum, — the increased distances must effectually counterpoise whatever infinitesimal effect they might exert, were it possible for them to act conjointly, when occupying points in their orbits nearest the earth.

In the same article (640), Herschel continues: "But . . . this diminution of the obliquity of the ecliptic will not go on beyond certain very moderate limits, after which (although in an immense period of ages, it being a compound cycle resulting from the joint action of all the planets), it will again increase, and thus oscillate backwards and forwards about a mean position, the extent of its deviation from one side to the other being less than 1° 21′." Here is an authoritive statement, ostensibly based on mathematical process: are there any means at hand of testing its real value?

In the year 1866, Sir C. Lyell, contemplating, in a new edition of his Principles of Geology, a thorough discussion of the causes of former changes in climate, wrote to Sir John Herschel, as competent astronomical authority, for precise information as to the possible extent of the supposed oscillation. In reply it was stated that the limit of 1° 21′, "as calculated by Laplace, is true as regards the last 100,000 years, yet if millions of years are taken into account, *it is conceivable* that the deviation may possibly be sometimes greater, and may even be found to extend as much as three or even four degrees on each side of the mean." * The state of uncertainty indicated by this extraordinary

* Prin. of Geol., vol. i. p. 293.

qualification of the original statement, and by the use of the phrase I have Italicized, is further attested by the sentence following the above citation, evidently suggested by, if not a literal quotation from, the astronomer's letter, which may fairly be construed as a virtual disclaimer of any positive knowledge on the subject. The sentence alluded to is as follows: "The questions entered into by Laplace and Leverrier respecting secular changes of the ecliptic relative to a fixed plane, and possible changes in the position of the earth's equator, must be the subject of laborious computations before astronomers will have decided what may be the extreme range of obliquity."

It is no impugnment of the infallible exactitude of just mathematical process based on proper elements, to say with Professor Huxley, when alluding to "the many cases in which the admitted accuracy of mathematical process is allowed to throw a wholly inadmissible appearance of authority over the results obtained by them;" that "mathematics may be compared to a mill of exquisite workmanship, which grinds you stuff of any degree of fineness; but nevertheless, what you get out depends upon what you put in; and as the grandest mill in the world will not extract wheat-flour from peascods, so pages of formulæ will not get a definite result out of loose data." *

In the matter in hand, the grinding process is confessedly far from complete. It has proceeded far enough, however to furnish samples by which to test the general quality of the results.

It will be observed that the original statement is adhered to, relative to the calculated amount of variation in obliquity for the last 100,000 years; that amount being 1° 21′. Now, at the present rate of progression, 48″ per century, which we know has prevailed for three or four thousand years, and probably for six thousand (for had the motion been slower than now, or not persistent, the ancients would hardly have detected it), a lapse of 100,000 years would carry it through about 13° 30′, instead of 1° 21′, the latter amount being accomplished in 10,150 years.

* Lay Sermons, Addresses, and Reviews, p. 249.

According to the table on page 176, showing the variations in the eccentricity of the terrestrial orbit for the last million years, such eccentricity has decreased from eight and one half to three millions of miles within the last 100,000 years. The extreme range of eccentricity, as has been stated, is something less than 14,000,000 of miles; therefore the decrease in that time would be about $\frac{5\frac{1}{2}}{14}$ of the whole possible extent. If variations in the eccentricity of the earth's orbit, and changes in the direction of the earth's axis of rotation, proceed from a common cause, and, consequently, are relatively proportional to each other, and, as stated by Herschel, 1° 21′, the assumed amount of diminution for the last 100,000 years is therefore $\frac{5\frac{1}{2}}{14}$ of the supposed whole extent of the polar oscillation, it cannot possibly "extend to three or even four degrees on each side of a common mean," but its whole value can be but little more than 3° 30′, and its extent, on each side of such common mean, but about 1° 45′, instead of three or four degrees.

These simple and practical tests serve to indicate the loose and irrelevant nature of the data and methods upon which the "stability of nature" theory is founded. Mathematical process that leads to indefinite results is absolutely valueless. The calculations of Laplace and Leverrier are exact and reliable in data, methods, and results, or they are not. From the foregoing considerations we seem entitled to infer that they are not. If the extreme amount of variation of obliquity may possibly vary from 1° 21′ to "as much as three or four degrees about a common mean," or six and eight degrees, nearly six times the former amount, and still accord with the calculations of Laplace and Leverrier, why may it not possibly vary even as much as six times the latter, or 48°? And further, if an oscillation of 90° is found to harmonize more closely with otherwise irreducible natural phenomena, what is to prevent our assuming that sum as the probable extent? Having arrived at this point, we find that the chances are largely against the probability of even this last amount, constituting the limit of the supposed oscillation; for the geological record indicates both coincidence with and perpendicularity of the ecliptic to the equator; and reckoning by degrees, the chances are as ninety to one against both of these positions being included in one quadrant of the circle. And if the motion may proceed for more than 90°, why may it not continue indefinitely?

Surely the results of mathematical processes, which, in their attainment, admit such an exercise of the imaginative faculty, — processes in which we are at liberty to *conceive* whatever the exigencies of favorite theories may seem to require, — cannot be permitted to constitute the sole obstacle to a grand, comprehensive, and rational generalization of the phenomena of nature.

Theorem of Lagrange. — Professor Mitchel, in advancing the theorem of Lagrange as a final solution of questions involving the extent of the supposed oscillatory movement of the earth under consideration, endeavors to show "how important the stability of the inclination of the earth's axis is to the well-being of the living and sentient beings now on the earth's surface;" a proposition undoubtedly true in a certain limited sense. The present conditions of life are essential to the well-being of the present inhabitants of the globe. But it is no less true that the conditions of life prevailing at some former period in the world's history, although differing widely from those which now obtain, were just as essential to the well-being of the forms of life then inhabiting the earth, as the present conditions are to the existing inhabitants. The error lies in assuming that the structure, and more especially the habits, of plants and animals must always remain the same, and, consequently, that the conditions of life cannot be subject to variation. But geology teaches us that in the past, radical changes of climate have occurred, involving essential modifications in the conditions of life; and palæontological science shows that, in the long run, nearly every form of life has experienced a continuous modification in structure. Analogy, and whatever evidence there is bearing on the subject, point to corresponding modifications of habit. Assuming, then, that the two classes of changes have proceeded in unison, and at an exceedingly slow rate, and every difficulty disappears. The original proposition holds good, not only at the present time and under the present status, but also at any other past era, whatever may have been the prevailing conditions of life incident to such era. It is erroneous, however, when applied to the whole extent of the cosmos, for that would imply that, since the creation of the world, there have been no climatic changes of sufficient importance to affect the conditions of life — a conclusion which carries with it the inference that, from the commencement of organic life, there has been no change in the structure or habits of the animals and plants inhabiting the earth.

Having started with this evidently erroneous proposition, mathematical aid is invoked to sustain it. As we proceed to the examination of the argument, let us bear in mind the substance of our last citation from Professor Huxley. Says Professor Mitchel, "Under the powerful and masterly analysis of Lagrange, this subject was completely exhausted, and a result reached which in the following proposition, guarantees the stability of the inclinations through all ages: — 'If the mass or weight of every planet be multiplied by the square root of its major axis, and this product be multiplied by the tangent of the angle of inclination of the plane of the planetary orbit to a fixed plane, and these products be added together, their sum will be constantly the same.'

"Now," continues Mitchel, "we will show hereafter, that the major axes remain nearly invariable, the masses of the planets are absolutely so, and hence the third factor of the product, *the tangent of the inclination*, can only vary within narrow limits, returning at the end of a vast cycle to the primitive value." *

On what I admit may prove to be a superficial view, it is difficult to perceive what it is in this proposition that constitutes the alleged guarantee of "the stability of the inclination." Suppose, in the case of one of the planets, — the Earth, for instance, — the inclination should be any other than the present value,— three, thirty, sixty, or even ninety, degrees, instead of 23° 27′, — the fact would not seem to invalidate the theorem: assuming its justness and applicability, it would only necessitate corresponding relative change in the inclination of one or more of the other planets; for the result, if it point to anything definite, only seems to indicate an invariable aggregate of inclination incident to the axes of all the planets. If, therefore, in

* Popular Astronomy, p. 306.

the same time it might take, at the present rate of progression, to bring the Earth's equator to coincidence with the ecliptic, diminishing the inclination from about 23° to 0°, the inclination of the axis of Mars to the plane of his orbit should increase proportionally, or from its present value, about 28° to 51°, or 28° plus 23°, the supposed invariable aggregate of inclination incident to all the planets, as far as the two were concerned, would be preserved, and they might continue the motion perpetually without detriment either to the theorem of Lagrange or the solar system.

But who is prepared to say that the permanence of the solar system is involved in the least in the manner in which one or all of the planets shall present their surfaces to the sun? If any degree of inclination is essential to the perpetuity of any given planet, it must be just as essential to that of each and all of the others. If an obliquity of about 23° is essential, in the case of the Earth, to the well-being of its inhabitants, and to the stability of nature in general, why is not the same obliquity essential in the case of Jupiter, or Saturn, or Mercury? If Saturn's axis of rotation is inclined at an angle of 63°, and that of Jupiter but 3°, why may not that of the Earth assume either position without disastrous consequences?

Table showing the Mean Diameter, and Eccentricity, of some of the Planetary Orbits, — Ratios of Eccentricity to Diameter, — and Obliquity of the Planes of their Orbits to their Equators.

NAMES.	Mean Diameter of Orbits in miles.	Eccentricity of Orbits in miles.	Approximate Ratios of Eccentricity of Orbit to Mean Diameter of Orbit.	Obliquity of Orbital Plane to Equator.
Mercury, . .	74,000,000	7,557,630	$\frac{1}{10}$	70°
Venus, . . .	136,000,000	473,100	$\frac{1}{811}$	75°
Earth, ...	190,000,000	1,597,325	$\frac{1}{114}$	23° 27′
Mars, ...	288,000,000	13,474,515	$\frac{1}{21}$	28° 42′
Jupiter, . . .	980,000,000	23,762,635	$\frac{1}{41}$	3° 4′ 5″
Saturn, . . .	1,800,000,000	50,958,399	$\frac{1}{35}$	63° 10′

By referring to the accompanying table, showing the mean diameter and eccentricity of the planetary orbits, ratios of eccentricity to diameter, and the obliquity of their orbital planes to their equators, it will be seen what a great difference exists in the inclination of the axes of the several planets. That of Jupiter is the least, being only a little in excess of 3°, and thence the values increase in the most indeterminate manner to 75°, the amount of inclination of the axis of Venus. It having been assumed that the axis of the Earth must be stationary or nearly so, it becomes necessary to suppose that those of all the other planets are so likewise, whatever their degree of inclination. In order, therefore, to retain the Earth, Mars, and Neptune in positions with regard to the Sun compatible with the existence of organic life on their surfaces, the remaining planets are doomed perpetually to remain in positions under which such life is held to be impossible.

In the absence of positive knowledge, it seems much more rational and consistent to reject the idea of a correlation subsisting between whatever motions may prevail among the planets analogous to that involved in the diminution of obliquity of the ecliptic, and assent to that only which leaves all of these bodies free to respond to the attractions and counter-attractions of whatever masses may exist in or approach their immediate vicinity. Under this view, instances might occur in which the determining causes of such motion would be entirely wanting, or where the effect of the same might be neutralized by other forces, leaving the axis of the planets stationary; others, where the various attractive influences might operate to produce an oscillatory or reciprocal motion; and others, as in the case of the Earth, where they would determine complete revolutions of the planetary masses; and no valid reason appears why we may not suppose all these resulting positions, whether stationary or varying, equally consonant, other necessary conditions concurring, with the development and continuance of organic life.

In leaving the problem, Mitchel continues, — " We are again indebted to Lagrange for the resolution of this most important of all the problems involving the stability of the solar system, who presents the final results as follows: — 'If the mass of each planet be multiplied by the square root of the major axis of its orbit, and this product by the square of the tangent of the inclination of the orbit to a fixed plane, and all these products be added together, their sum will be constantly the same, no matter what variations exist in the system.'

" The mass or weight of each planet," says Mitchel, " is

invariable, while the loss or gain in the values of the major axes is always counterpoised by the gain or loss in the inclinations of the orbits, and thus in the long run, in cycles of vast periods, a complete restoration of the major axes is fully accomplished, and the system in this particular returns to its normal condition. The entire problem," he adds, "is very complex. With difficulties so extraordinary; with complications and complexities mutually extending to each other, involving movements so slow as to require ages for their completion, it is a matter of amazement that the human mind has achieved complete success in the resolution of this grand problem." *

These congratulations appear to be somewhat premature; for, as before stated, Lyell informs us, on no less authority than that of Sir John Herschel, that these questions "must be the subject of laborious computations" before this "complete success" can be achieved. Admitting, however, that what appears to be a complete resolution has been effected, if the process involved such extraordinary and diverse complications and difficulties, who shall pronounce upon the exact propriety of the methods employed, supposing the basis to be just and the elements all applicable to the question? Or allowing the methods to be proper and correct, and the basis of the solution just, are all of the multitude of complex and diverse elements that enter into the problem pertinent? A single error in any part necessarily vitiates the whole performance. The last recited proposition of the French geometer is based on the assumption — one which we seem to encounter in all quarters — of correlation between changes in the inclination of the planetary axes of rotation, and variations in the eccentricity of the orbits; both, in the case of the Earth, being assumed to proceed from a common cause, viz., the joint action of all the planets, principally the nearer and large ones, Jupiter and Saturn, Venus and Mars; and it is asserted, that through the operation of this common cause, the loss or gain in eccentricity is counterpoised by the gain or loss in inclination. If this were actually the case, as the inclination of the Earth's axis is known to be now diminishing at the rate of 48″ per century, the eccentricity

* Popular Astronomy, p. 314.

of its orbit ought to exhibit a correspondingly proportional increase; while, on the contrary, it is well known to be decreasing in value, the terrestrial orbit at present approaching or tending towards a circular form.

This fact, together with the considerations previously advanced, goes far to show the erroneous nature of this assumption, in the case of the Earth at least. But, allowing it to be so far true, who has determined the fact that there is any variation in the degrees of inclination of the axes of the other planets, or in the eccentricity of their orbits? If this is a satisfactorily ascertained fact, and one or both classes of variations do actually occur, are they the product of the same agency that determines the terrestrial motions? It can hardly be supposed that the producing cause can be precisely the same in all cases, but must vary in each instance according to the proximity and volume of the neighboring planetary masses. This being the case, or, if the supposed variations are consequent on separate and distinct causes, the ratios of these variations would, most likely, or rather must, differ in each and every instance, and thus vitiate the supposed solution, which, at best, is purely theoretical, and cannot be submitted to any practical test.

Again, if a correlation subsisted between changes in the inclinations of the planetary axes of rotation and variations in the eccentricity of their orbits in consequence of both proceeding from a common cause, such correlation, it would seem, must be evinced in the case of each planet by a relative correspondence between the amounts of each. The table, however, shows that no such correspondence exists. The eccentricity of the Earth's orbit, as before stated, is now diminishing, and so likewise is the inclination of its axis of rotation. It follows, therefore, that if there is a relation between the two, and they are both results of a common cause, that cause, in the case of each planet, operates to increase or diminish both at the same time. A large eccentricity in the orbit of any given planet would then be accompanied by a large inclination of its axis, and a small eccentricity by a correspondingly small inclination. Such is not the case. Venus, whose orbit approaches nearer to a circular form than that of any other planet, the eccentricity being to the mean diameter of orbit only as about 1 to 311, exhibits a very large inclination of axis — about 75°; while Mercury, with nearly as great an inclination (70°), shows the very largest degree of orbital eccentricity, the same being equal to one-tenth of the mean diameter of its orbit. Saturn, whose axis is inclined from perpendicularity with its orbital plane at an angle of 63° 10′, has also a large eccentricity of orbit; about one thirty-fifth of its mean diameter; while Jupiter, whose orbit exhibits about the same degree of eccentricity as Saturn's, — about as 1 to 41, — and whose inclination, according to the hypothesis, ought to be about 54° to accord with that of Saturn, shows, on the contrary, the least inclination of all the planets, amounting to but little more than 3°. The Earth and Mars, with nearly the same inclination of axis, — that of the former being about 23°, and that of the latter 28°, — present, the one, next to the highest, and the other next to the lowest degree of orbital eccentricity of all the planets. What more conclusive evidence can be adduced to show that no correspondence whatever subsists between these two planetary elements?

In an astronomical point of view, and aside from the positive evidence furnished by the geological record, we are justified in the conclusion that variation in the eccentricity of the Earth's orbit, and change in the direction of the terrestrial axis of rotation, are due to separate and distinct causes acting independently of each other. If, as is undoubtedly the case, variations in eccentricity are to be ascribed to the attractive influence of the nearer and larger planets, — Jupiter and Saturn, Venus and Mars, — variations in obliquity must be due to some other cause. That which we have assigned seems the most obvious and natural, and until some more valid reason can be found as a warrant for its rejection, than that it involves a continuous rotatory motion of the Earth, pointing in time to a period characterized by complete uniformity of seasons, we need not hesitate to accept it as the true theory.

Nothing, perhaps, can be plainer to an unbiased mind, than that the present received theory of diminution of obliquity is the result of attempts to conform physical phenomena to preconceived erroneous notions of a hypothetical stability of nature, the truth of revelation being involved, as appears to be supposed, in such stability. Not only must any considerable degree of variation in the position of the Earth's axis prove generally disastrous to the entire solar system, by disturbing the delicately adjusted balance of antagonistic forces upon which depend the undeviating order and regularity characterizing the movements of the planetary masses, but also the effect of such variation upon the Earth would be, if tending to greater obliquity, to render its surface uninhabitable to organic life, or, in the opposite direction, by producing a universal uniformity of seasons, abrogate the promise made to man, that as long as the Earth shall continue, day and night, summer and winter, &c., shall not cease.

To the first of these objections we have opposed the common-sense view that, the form of the Earth being globular, the attractive influence of other planetary bodies upon it, and its own reciprocal attraction upon them, must be the same in every position it is possible for it to assume; to the second it has been urged that the pliant nature of organic life is competent, by means of self-modification, to adapt it to all and every change in the conditions of life possible in the premises. To the third, which seems more properly a subject of theological than scientific discussion, only a few brief observations will now be advanced.

Says Professor Hind, in this connection,* "The change of obliquity is a phenomenon in which we are concerned only as astronomers, since it can never become sufficiently great to produce any sensible alteration of climate on the Earth's surface. A consideration of this remarkable astronomical fact cannot but remind us of the promise made to man after the deluge, that, 'while the earth remaineth, seed-time and harvest, and cold and heat, and summer and winter, and day and night shall not cease.' The perturbation of obliquity, consisting merely of an oscillatory motion of the plane of the ecliptic,† will not permit of its ever

* Solar System, p. 46.

† It may be proper here to caution the reader against accepting a shadow for the substance — an appearance for the reality. The sun *appears* to revolve about the earth; but such appearance is only an effect of a real motion of the earth — its diurnal rotation. So, whatever its nature and extent, the motion of the ecliptic referred to is a mere hypothetical appearance, consequent on another real motion of the earth, whatever may be the nature and extent of such motion, transverse or across the diurnal movement.

becoming very great or very small, is an astronomical discovery in perfect unison with the declaration made to Noah, and explains how effectually the Creator had ordained the means for carrying out his promise."

In relation to this branch of our subject, it may be observed, that, differing only in degree, every phase of the phenomena of day and night, summer and winter, cold and heat, and seed-time and harvest, that must follow the entire series of changes dependent upon the transverse rotation of the earth, have their exact counterparts on various portions of the earth's surface at the present time. The equal division of the year into day and night, by which every portion of the earth's surface enjoys the same absolute annual proportion of each, would continue throughout the whole transverse revolution the same as now. Days of every degree of length, from those determined by the momentary appearance of the sun upon the horizon, to those of more than five months' duration, with nights relatively proportional, are familiar to explorers in high latitudes, who might, were it possible for them to penetrate to either of the poles, even now behold the phenomenon alluded to in ancient tradition, of a year divided into one day and one night, each of six months' duration. If these facts are not in contravention of the promise, a change from whatever variations in the relative length of day and night may now prevail in the United States, Middle Europe, and Syria, to those above indicated, certainly would not be. There are also insulated portions of the earth's surface where various local causes unite to produce very equable climatic conditions, and where the distinctions of seed-time and harvest, summer and winter, as familiar to the inhabitants of the regions above specified, are quite unknown. Who, also, shall discover other than a hypothetical winter at the equator, or summer near the poles? Who on the ice-encircled shores of Smith's Sound, or among the glaciers and snows of Grinnell's Land, divides the year into seed-time and harvest?

It cannot be supposed that these instances, so much at variance with our local ideas, militate against or are antagonistic to the divine promise. If not, neither do whatever analogous conditions may be consequent upon the transverse revolution of the earth. We may also consider the period of equatorial coincidence with the ecliptic, or era characterized by uniformity of seasons, as a mathematical point; so that no sooner does obliquity cease upon one side, than it immediately recommences on the other.

The permanence of the solar system, the existence and well-being of organic life upon the earth, the validity of revelation, and, in a word, the stability of nature, having been assumed to depend upon the practical immobility of the earth's axis, and having accorded to this most erroneous assumption the rank of a fundamental truth, it became necessary to bring the results of scientific investigation into conformity therewith. As long as the opinion was universally prevalent that the past, and perhaps future duration of the earth was to be limited to a term of time which might be included in a single decade of thousands of years, no objection could appear to the simple and natural hypothesis that referred the diminution of obliquity to the unequal attraction of the moon upon a prominent irregularity in the form of the earth. But when geologists, from a more intimate acquaintance with the terrestrial surface, came to the contemplation of periods in the earth's history which must have embraced hundreds of thousands, if not millions of years, and learned, further, that life in various forms had continued in existence throughout those vast periods, these considerations, together with the fact that no general disturbance in the order of the solar system had arisen during the whole time, made it appear evident, reasoning from the premises assumed, that the terrestrial motion upon which diminishing obliquity is consequent, could not possibly be the result of lunar attraction; for if so, as the movement could never become stationary nor increase, — i. e., assume an oscillatory character, without power extrinsic to the solar system, — it must, perforce, proceed indefinitely in one direction; a contingency not to be thought of. The lunar agency was, therefore, set aside, and the "joint action of all the planets" substituted. Nothing could be easier than to assert the planetary agency, and nothing more difficult, perhaps, than to prove it. But something must be done in that direction, and the "Mathematical Mill" is set in operation, with the inconclusive and unsatisfactory results already noted.

We are informed at this juncture, however, that the

whole question is one that can interest us only as astronomers, since the movement can never proceed far enough on either side of the mean to affect the well-being of the inhabitants of the earth, by producing sensible variations in climate; and we are left, naturally enough, to infer that even if inapposite and erroneous methods of solution have been employed, the result may as well be accepted as a finality, whatever its character, the whole subject being of no practical importance.

From no point of view can any true student of nature assent to such a disposal of it. Most assuredly, then, those who behold, as hinging on this motion of the earth, some of the most important processes of nature, affecting largely the relations of the organic to the inorganic world, cannot attach the least value whatever to conclusions reached through complicated and doubtful methods, based on numerous complex and obviously irrelevant data, more especially when such conclusions form the only obstacle in the way of an extensive and harmonious generalization of hitherto irreducible natural phenomena.

A just natural system must always be harmonious, consistent, and, above all, simple and complete. It is impossible that it should be otherwise. On the other hand, an erroneous one must always bear the impress of complexity, inconsistency, and incompleteness. We cannot but recall to mind, in this connection, those features of the Ptolemaic system of astronomy, analogous to some of those characterizing the theory which endeavors to make the stability of nature — the permanence of the solar system, and the existence and well-being of the earth's inhabitants — dependent upon the continuance of the present status. The Ptolemaic system taught that the earth was the centre of the solar system; that it was at rest, and that all the heavenly bodies revolved about it as a common centre. The philosophers of the times imagined, doubtless, that the stability of nature was involved in its immobility, as the pivot of the universe. Geometers invented circles and curves of almost every conceivable description to systematize the apparent motions of the sun and moon, planets, comets, and stars. So successful were they, to all appearance, that although Pythagoras, five hundred years before the Christian era, had privately taught his disciples the true theory of the solar system, that of Ptolemy was for fourteen hundred years universally acknowledged as the true one.

During all this time, however, there was one open, palpable, and fatal objection to it. The utterly impossible and inconceivable velocity, which, according to it, the celestial sphere must have attained in order to accomplish a revolution about the earth in each twenty-four hours, ought, as it now seems, to have been sufficient to cause the rejection of the whole system, involving as it did, as a necessary element, such an absurdity. But it did not; and the Ptolemaic system of astronomy became so firmly established and so deeply ingrafted upon all the prevailing modes of belief, religious and secular, that Copernicus, when giving to the world the true system in his work "On the Celestial Revolutions," in order not to shock too rudely received opinions and prejudices, found it expedient to present it under the form of an hypothesis. "Astronomers," said he, in his dedication to Paul III., "being permitted to imagine circles to explain the motion of the stars, I thought myself equally entitled to examine if the supposition of the motion of the earth would render the theory of these appearances more exact and simple."

Not even the harmonious simplicity which the revived system substituted for the complicated disorder of that it was destined to supersede, could serve to shield it from the attacks of over-zealous defenders of revelation. That it was in opposition to their ideas of what constituted the literal meaning of certain passages of Scripture, was sufficient not only to excite violent hostility, but led also to the devising of systems still more complicated, if possible, than that of Ptolemy, like that of Tycho de Brahe, curiously compounded of the true and false, by means of which it was hoped to reconcile the imaginary discrepancies. The labor, however, was in vain. Objections and false theories have passed away, and religion and the Copernican system now proceed harmoniously together, hand in hand.

Among the points of resemblance between the old astronomical theory and that modern system which assumes that the permanence or stability of nature is dependent on the practical immobility of the earth with regard to the sun, in the sense employed in this work, is the fact that both have received general assent in the face of an unavoidable and fatal objection. The solar influence being invariable, we know from our own experience, that certain phenomena must follow the presence and direct action of the sun's rays, and also that certain other effects must result from their absence or indirect action. If, therefore, the earth and sun have always held their present relative positions, one of these classes of phenomena must invariably be confined to the neighborhood of the equator, and the other to the vicinity of the poles. As it is now an unquestionable fact that proofs of the actual presence and direct action of the sun abound at or near the poles, and also that an epoch in the past bearing the impress of its absence or indirect action, may be traced within the tropics, it is as absurd to insist that the sun has always, throughout all past time, followed its present path in the heavens as it was for our ancestors to cling to the erroneous idea that all the planetary and stellar bodies in the universe performed a complete revolution about the earth within each twenty-four hours.

Another point of resemblance is, that each has received a factitious support from what has assumed to be reliable and just mathematical process. If the principles of this, properly, most exact of all the sciences, are susceptible of an amount of distortion sufficient to afford a basis for the astronomical system of Ptolemy, it need occasion no surprise if they are found pliant enough to furnish an imaginary foundation for any other equally absurd theory.

Another analogy between the two is found in the fact that, on a superficial view, the opposed true systems present, apparently, points of antagonism to the literal sense of the Bible. But has true religion, in the former instance, experienced any permanent detriment from such erroneous impressions? Are not the heavens still telling the glory of God, and doth not the firmament still show His handiwork? Do the heavenly hosts speak a language less eloquent to our ears, and are the lessons they teach us of the infinite power and wisdom of the Creator less impressive upon our hearts as we become better and better able, through a more perfect knowledge, to comprehend them?

Our highest conceptions of the character and attributes of God must come to us through the contemplation of the marvellous works of His hands; and true science, as the analyst and demonstrator of those works, having in the former instance left no impediment in the way of genuine religion, so we may safely conclude, that whatever apprehensions of this nature may, in the present, be at first entertained, their groundless nature will soon become apparent.

CHAPTER VII.

GENERAL SUMMARY AND CONCLUSION.

STATEMENT OF THE THEORY OF THE TRANSVERSE ROTATION OF THE EARTH. — POSITIONS ASSUMED BY THE EARTH IN THE COURSE OF SUCH ROTATION, AND THEIR EFFECT. — RECAPITULATION OF PROMINENT FACTS, AND FURTHER OBSERVATIONS THEREON.

THE sun is the only source from whence the earth derives those supplies of light and heat which constitute it the abode of organic life.

In the absence of the light of the sun, the earth would receive a very small amount of light from the fixed stars; and without its heat, limited amounts of the same might occasionally be produced at the earth's surface, through mechanical and chemical agencies. Under no conditions, however, can the quantities of light and heat derivable from these sources be supposed sufficient for the development and sustenance of any form of organic existence.

The sun, as a source of light and heat, is an unchanging body; and the effect of its rays, at any given distance, and under like conditions, is always the same.

The earth, presenting continually the same proportion of its surface to the sun, and preserving an invariable mean distance from the same, the mean annual amount of light and heat received by the earth from the sun must continue invariable. Practically this is found to be the case, judging from the observed effect of greater or less distance of the earth from the sun, resulting from the elongated form of the terrestrial orbit; the tardy operation of the heat-distributing agents so diffusing the perihelion excess of heat over the whole year, that if ever felt at one time more than at another, it only becomes sensible six months after its transmission to the earth. This fact seems to indicate a close approximation to an invariable *daily* amount. According to Sir John Herschel, however, a variation to the extent of three one-thousandths of the whole annual aggregate is possible under the greatest extremes of eccentricity the earth's orbit can ever attain; a quantity so small, as he remarks, that it may be omitted in inquiries of this nature.

The annual amounts of heat received by the earth from the sun have, therefore, always been approximately, if not absolutely, the same.

But this annual aggregate of solar heat is found to be very unequally distributed over the surface of the planet. From the globular form of the earth, the rays of the sun meet its convex surface at various degrees of inclination; the same being perpendicular to the plane of the horizon at the centre of the earth's disk, as presented to the sun, and varying thence through every degree of angularity, until they become parallel with such plane at the circumference. The heating power of the solar ray is found to be proportional to the angle of incidence it makes with the earth's surface.

If, then, the sun is the only source of heat in quantity adequate to affect the terrestrial climate, and the supply it furnishes is invariable, it follows that as long as the mean annual inclination of the sun's rays at any place remains the same, the mean annual temperature of such place must also remain the same, — saving only the comparatively trivial effect of *modified* convection. Primarily, inclination determines climate. The mean annual inclination of the sun's rays at any place is immediately dependent on the earth's position with regard to the sun, or on the inclination of the earth's axis. A radical change in the mean annual temperature of any given latitude is, therefore, impossible, without a corresponding variation in the position of the earth's axis, and a consequent change in the inclination of the solar rays in that latitude. If no such variation in the relative position of the earth and sun has ever occurred, — if the earth's axis of rotation has always been stationary, or approximately so, — no appreciable deviation from the present climatal status, or mean annual temperature, of the several terrestrial zones can ever have occurred, such a deviation being in the nature of things impossible, as having no adequate cause. The annual amount of heat received by the earth being always the same, if we assume this amount to have been always distributed over the earth's surface in the same manner and by the same method as at present, each part thereof receiving perpetually the same proportion of the annual aggregate it now does, the geological record must, and would, indicate throughout all past time a torrid climate at and near the equator, a frigid climate at and near the poles, and a temperate climate in intermediate latitudes. Interchanges of heat and cold between the extreme points could not, through the agency of aerial and marine currents, proceed at a more rapid rate at any former period than they do now, under similar circumstances; and possible changes in the distribution of land and sea, and in the elevation of land, could produce only local and no general effect. Nothing can be more certain, then, than that, under the present status, the present climatal conditions must inevitably ensue; and on the contrary, nothing can be more absurd than the supposition that under the same, the earth and sun holding their present relative positions, anything analogous to a torrid climate can obtain at the poles, or that the glacial phenomena of polar latitudes can ensue at the equator. This assumption plainly involves such changes in the inherent properties of the solar rays as shall produce a reversal of the thermal effect of greater or less inclination; for under no other contingency could it be possible for the solar influence — the sun pursuing its present apparent path in the heavens — to so operate as to produce glacial phenomena at the equator, or a florid vegetation at the poles.

Geological research develops no fact with greater precision and certainty, than that, in remote periods of the earth's past history, these, to us abnormal climatal conditions, have prevailed. As far as modern effort has succeeded in penetrating in the direction of the poles, every evidence is met with, that can reasonably be expected, of not only one, but a succession of epochs marked by an abounding animal and vegetable life; such life being of a character as evidently to require for its development an amount of heat by far exceeding that now prevailing there. On the other hand, the observations of the late Professor Agassiz, in tropical South America, furnish the most positive evidence of the former prevalence in those latitudes of a climate under which extensive glacial processes could go on, at moderate elevations at least, if not, indeed, at ordinary levels. Middle latitudes, too, are found geologically to exhibit every indication of former violent extremes of heat and cold, as compared with those of the present time; and a gradual transition is also apparent from an anterior period of uniform warmth or equality of seasons to one of apparent excessive cold, or, rather, of great annnal exhibits of heat, and cold (the traces of the cold being of a more durable nature than those of warmth, this being the reason of the supposititious excess of cold of the Glacial period), and thenceforward a like apparent decrease of cold, or actual decrease in the inequality of seasons, to the climatic conditions of the present time. Numerous instances also appear in the earth's history, of which the Miocene and Carboniferous periods exhibit prominent examples, of intervals of time in which the radical difference now observable between the climate of high and low latitudes became, to a large extent, neutralized, and a climatal analogy prevailed over the whole northern hemisphere, from east to west, and north to south, and, as we have reason to believe, from the equator to the pole; certainly from the West Indian Islands to places situated within the arctic circle.

These, in brief, are our facts. They admit neither of doubt nor cavil. What do they indicate?

Obviously, if the sun is the only source of heat determining the terrestrial climate, and if the solar influence is in quantity and effect invariable, and its exhibition under like conditions must inevitably produce like results, the above facts, as indicating effects of the solar influence the most unlike possible to those which now follow its exhibition, clearly demonstrate a corresponding diversity of conditions attendent on its exhibition in those former eras. Warmth at the poles not only indicates, but demonstrates, the solar presence at the poles. Cold at the equator equally demonstrates the solar absence at the equator. An analogy of climate throughout all latitudes demonstrates an entirely different distribution of solar heat from that now prevailing. If the annual amount of heat received by the earth from the sun is constant, whenever polar areas shall receive a larger proportion of this aggregate than they now do, equatorial areas must receive a proportionally diminished quantity. Changes in the mode of distributing the annual amount of solar heat over the surface of the earth, can be effected in no possible way other than by changes in the position of the earth with regard to the sun, consequent on variation in the direction of the earth's axis of rotation. Sudden changes of this nature and their inevitable consequences being alike contrary to the analogies of creation and the evidence furnished by the geological record, they and the theories depending upon them may be dismissed as impossible. A slow, constant, and uniform motion of the earth at a right angle to its diurnal revolution would be perfectly competent to effect these climatal changes, and at the same time, if slow enough, would afford opportunity for the organisms inhabiting it to adapt themselves, through their innate plasticity of constitution, to the continual change in the conditions of life. A movement of the earth of this nature is now in progress, and, as far as our direct knowledge extends, always has been. Its continuity is involved in the facts recited. There appears, therefore, no logical escape from the natural and legitimate conclusion to which our inquiries have conducted us, which may be stated as follows: —

The earth, in addition to its diurnal revolution, has another proper rotatory motion, transverse, or across the diurnal rotation; the same determining, in stated periods, complete revolutions of the planet.

The stability of nature, then, as regards the earth, does not consist in its practical immobility in one particular direction, but rather in the uniformity and constancy of each and all of its proper motions; and there is no less reason to suppose that the one of them which has the effect to induce those successive cycles of climatal change, with all their varied and important results in the terrestrial economy to which our attention has been, in this discussion, more particularly directed, is as persistent and uniform from century to century, and from epoch to epoch, as is that which produces the alternations of day and night. A geological period must invariably represent an absolute, determinable lapse of time. Not but that the transverse rotation may proceed at a more rapid rate during some portions of such revolution than during others; like the annual motion, in which the earth moves with greater velocity in some parts of its orbit than in others, although the mean is strictly invariable. Those elements which determine the value in time of any one year, determine also the length of every other year; and the elements upon which depends the duration of one great geological epoch control that of all the others.

If the transverse motion of the earth proceeds from the cause we have assigned, viz., the attraction of the moon on the excess of matter at the earth's equator, we have an element that obviously may more or less affect its velocity, if the word may properly be used in connection with a movement so slow.

It is evident that as the solar heat becomes gradually concentrated towards the equator in consequence of diminishing obliquity, the permanent superficial accumulation of ice and snow at the poles must be continually augmenting, not only in thickness, but must, moreover, continue to extend further and further southward. It is impossible now to say how far in this direction the permanent glacial deposits of the period of equatorial coincidence with the ecliptic would go; still it can hardly be doubted but that a proportion at least of the more marked traces of glacial action in northern temperate latitudes are properly referable to this time. It seems possible, therefore, that deposits of this character may thus accumulate to such an extent as to slightly diminish the effect of oblato-spheroidicity. The effect of the lunar attraction would, therefore, be least, in this respect, at the time of coincidence of the earth's axis with the ecliptic, when, in consequence of the more equal distribution of the solar influence over the terrestrial surface, no *surface* deposits of ice and snow would remain unliquefied throughout the year in any latitude; and thus the transverse rotation may be supposed to proceed faster than the mean in one instance and slower in the other.

Whether we assume the above cause adequate to the effect, or, on the other hand, regard the general form of the earth, as determined by the velocity of the diurnal motion, absolutely invariable (a proposition strictly true of the fluid portion of it), until further data are obtained than we now possess, 48″ per century may be regarded as the mean rate of the transverse movement of the earth. A geological period, or half of a complete transverse revolution, would thus require for its accomplishment 1,350,000 years, as previously stated.

Effect of the Transverse Motion of the Earth on the apparent Motion of the Sun. — In order to obtain a general idea of the climatic effects of the various positions the earth assumes, with regard to the sun, in the progress of the transverse revolution, we may, as the most convenient starting-point, perhaps, take the last period of equatorial coincidence with the ecliptic, which, according to our method of computation, was some 11,740 centuries ago. It constituted the intermediate stage of a geological period of subsidence or repose, as contradistinguished from the glacial interval or period of change, which separated the present or Pliocene epoch from its immediate predecessor, the Miocene. The days and nights were then of equal length all over the earth, except at the poles, where, throughout the year, the sun would appear to describe circles about the horizon coincident with its plane. There would be no variety of seasons as determined by the present motion of the sun north and south of the equator, that luminary following daily the same apparent path through the heavens, such path, at the earth's equator, being perpendicular to the horizon, and coinciding with the plane of the celestial equator. The solar rays would therefore be vertical each day at meridian throughout the year, becoming inclined at points north or south of the equator, the degree of inclination at any given point being indicated by the latitude of such point. Thus, in latitude 30° north or south, where, as in all others, the sun would rise invariably in the east, and precisely at six o'clock, its rays at meridian would exhibit an inclination of 30° from the vertical. At latitude 45° the meridian inclination would be forty-five degrees, and so on, until at the poles the sun would not rise above the horizon during the day of the rest of the earth, or fall below it at night, but would appear to revolve continually about the horizon, one half of its disk above and the other below the same. The general result would be continuous summer at the equator, continuous spring in middle latitudes, and continuous winter at the poles. The climate of the tropics, for the reason heretofore given, would probably not vary to any very great extent from that of the present time. The temperature of middle latitudes, however, might become depressed considerably below the present mean by permanent surface accumulations of snow and ice, which would remain throughout the year in latitudes much lower than at present.

The condition of continual winter at the poles, summer at the equator, and spring in intermediate latitudes, would have no effect to disturb the superficial repose of this era. In the warm latitudes there would be no more, certainly, and probably less, change in the position of the material composing the earth's surface than there is at present; and towards the poles, the retardation and perhaps total arrest of meteorological process, and the continuous ice-bound condition of the land, would effectually prevent change in the aspect of its surface. What we have heretofore said in relation to the condition of the whole earth during the imaginary "cosmic winter," appears to be really applicable to high temperate and polar latitudes at this time, which, we believe, may with propriety be denominated the true Polar Glacial period. Even the transitions to and from this state of inert frigidity — necessarily extremely slow processes — would not in the least disturb the condition of superficial repose. Assuming the earth, in this era, to have been peopled by beings of sufficient intelligence to transmit the results of observation from age to age, after the manner of the present time, through the accumulated records of thousands and tens of thousands of years, they would be unable to detect any appreciable amount of change, either climatic or superficial, and, with much more apparent reason than we now have, might conclude that the various conditions under which they lived were, of necessity, absolutely essential to the stability of nature and to the perpetuation of life.

It has been inferred from certain observed facts that the whole earth's surface from the poles to an undetermined distance towards the equator, but somewhere between the 40th and 50th parallels of latitude, perhaps, was once buried beneath superincumbent masses of ice, thousands of feet in thickness. If these polar ice-caps, as they have been called, ever existed, it is this era to which they are properly referable, and not to the so-called Post-Pliocene Glacial period, although it is more than probable that the phenomena peculiar to each of these totally distinct ice-periods have been more or less confounded with, and mistaken for, those of the other.

While the status just described would remain apparently unchanged for an immensely long period of time, the transverse motion of the earth, as the centuries rolled by, would generate an annual movement or deviation of the sun north and south of the equatorial plane in which it appeared to revolve while that plane was coincident with that of the earth's orbit, the deviation increasing by 48″ on each side every one hundred years. At the same time the sun would begin to exhibit at the poles a gradual annual rise above and dropping below the horizon, the deviations from its plane increasing by 48″ per century. These centennial amounts of variation accumulating for sixteen hundred eighty-seven and a half centuries, the time consumed in effecting one fourth of a quarter of the transverse revolution, would bring the earth into a position, with regard to the sun, analogous to that which it now occupies, its axis being then inclined at an angle of 22° 30′. This position will again be attained by the earth in a little more than seventy centuries from the present time. The peculiarities of climate and diversity of seasons of this era approach so nearly to those of the present time that no description of them is necessary.

The inclination continuing to increase at the same rate, in another period of sixteen hundred eighty-seven and a half centuries it will have reached a value of 45°; and at this date we have located the commencement of the Pliocene-Glacial period. The interval of time above referred to as preceding it, may properly be considered as a transitional period, leading from the anterior era of repose to the era of change, the increasing contrasts of heat and cold characterizing the seasons having begun even then to effect some considerable superficial change.

At this time, with the terrestrial axis inclined at an angle of 45°, the annual motion of the sun north and south of the equator, extending to 45° each side of the same, would produce in equatorial localities considerable variety of climate, each year being marked by two seasons of nearly the usual degree of warmth, alternating with two cooler seasons; the sun's rays at meridian varying from the vertical to an inclination of 45° twice in each year.

In latitudes 45° north and south, climate would be determined by an annual motion of the sun from the zenith at the summer solstice, when the longest day would consist of twenty-four hours, the sun being in the zenith at meridian, and on the polar horizon at midnight, to 90° from the same at the winter solstice, when the sun would just become visible at noon on the horizon in the direction of the equator. The day would therefore be 0, and the night twenty-four hours, each year having a day and night of that length respectively. The heat of midsummer must have been about the same as that now experienced within the tropics, while the winters would be arctic in severity. At the poles, the sun, after a six months' absence, descending in that time 45° below the horizon, and returning thereto, would make his appearance above it, describing complete circles about it, and gradually ascending in spiral circles until he attained an altitude of 45°. The descent would be accomplished in the same manner, the ascent from and the return to the horizon occupying the remaining six months of the year. The sun, during the long polar day, rising to nearly twice the height it now does, may easily be supposed capable of developing and supporting a considerable variety of vegetation, and many more forms of animal life than now flourish there; and types of the human family analogous to the Esquimaux tribes of the present time would, no doubt, be able to subsist within every portion of the earth's surface now included in the polar circles.

The transverse motion of the earth continuing, the intermediate point between an inclination of 45° and coincidence of the earth's axis with the ecliptic would be reached in another period of sixteen hundred eighty-seven and a half centuries. The inclination of the equator to the plane of the terrestrial orbit will now amount to 67° 30′, and it is this era that probably constitutes the most rigorous position of the Glacial period in intermediate latitudes.

The annual course of the sun at the equator will now extend 67° 30′ on each side of the same, approaching at its maximum declination north or south to within 22° 30′ of the horizon, thus considerably intensifying the semi-annual periods of cold. At the forty-fifth parallels, the annually recurring night of twenty-four hours' duration, characterizing the preceding era, will have increased in length to about six weeks, and the long day, in which the sun would appear to ascend in oblique spiral circles to within 22° 30′ of the celestial pole, would be of the same length. At the pole, the sun in its spiral ascent would attain an altitude of 67° 30′, approaching within 22° 30′ of the zenith, or the point in the heavens constituting at that time the celestial pole. The annual mean of temperature at the poles must now make some approximation to that of equatorial regions, although the distribution of their respective shares of the solar heat over the year would be very unequally effected.

After the lapse of another of these periods, the earth's axis will have become coincident with the ecliptic, the equator becoming perpendicular to the same. This position of the earth is that to which the traditions and sacred writings of the ancient Eastern nations refer. This era, and another of equal length immediately succeeding it, have been considered as constituting, in middle latitudes, the Inter-Glacial period. As we have had occasion before to state, the apparent course of the sun will now extend, north and south, 90° from the equator, to an observer there appearing to touch the northern and southern horizons at intervals of six months, dividing thus the year into two summers and two winters. At latitude 45° the year would now be divided into a night, or winter, of three months' duration, determined by the absence of the sun below the horizon,—three months of alternate day and night,—three months of continuous day, during which the sun in its oblique spiral ascent would reach the celestial pole, and then another three months' interval of day and night. At the poles, an unvarying day, or summer, of six months, and a night, or winter, of equal duration, would make up the year.

It will be observed that the earth, during these four eras, amounting in the aggregate to six thousand seven hundred and fifty centuries, has performed one fourth of a complete transverse revolution, and has experienced every variety of effect that can possibly result from change of position with regard to the sun. The first division of the quarter revolution was the last half of an antecedent period of repose; the next, an interval of transition from the period of repose to the glacial interval or era of change; the next, the glacial interval, and after that the first half of the Inter-Glacial period. The next quarter revolution includes, in reverse order, the same round of change, viz., the last half of the inter-glacial period, the glacial, the transitionary, including the present time, and, the world continuing, an era in the future comprising the first half of another epoch of repose, at the end of which the earth will have returned to the position with regard to the sun from whence we set out. The poles, however, would be reversed, a whole revolution being required to accomplish a complete restoration.

The half revolution described may be supposed to include all the members of a geological period, although not a complete one, as the true initial point of such periods must be the beginning of a glacial interval, and its close the end of an interval of repose.

There is, therefore, no stationary or normal terrestrial status anywhere in the earth's history, to which the many and various highly contrasted states and changes disclosed by the geological record can be compared as abnormal departures therefrom. The natural condition is one of continual change, of unending variation, affecting alike climate, the physical conformation of the earth's surface, and the organisms which inhabit it. Climatal changes, like the causes which produce them, are reduceable to determinate cycles, identical one with another. The climate of the earth at any given point, in any one of its transverse revolutions, is identical with that of the same stage of any other of such revolutions. Not so, however, with the changes in the earth's crust, and those of its living inhabitants. The round of variation as affecting these, instead of returning in stated periods into itself, shows, on the contrary, a uniform tendency towards progression and improvement. Each recurring glacial epoch leaves the terrestrial surface in a condition to support higher and higher types of animate and inanimate life, while, at the same time, the operation of immutable laws is continually evolving those higher and higher types, and adapting them to their improved habitation.

From the first dawn of life upon the earth down to the present time, the geological record shows an ever-accelerating progression in organic development. Who, then, in the face of this great fact, shall dare assert that the acme of progress has been attained, and that man, as the highest type of organic existence, has reached his ultimate goal, and must either remain essentially in his present estate, or retrograde?

A broader and more hopeful, more consistent and rational view is that which, ceasing to regard the present in any respect as a standard to which all other eras are to be made to conform, looks upon it rather as a mere link in that grand chain of progress which seems tending slowly, but surely, to bring our planet into the condition of an earthly paradise, and mankind morally, intellectually, and physically, into fit inhabitants for such an abode.

Recapitulation of Prominent Facts, and Observations thereon.

Within historical times there has been no appreciable absolute change in the climate of those parts of the earth's surface inhabited by civilized races of men. This fact is in strict accordance with our assigned cause of change, which, with its effects, proceeds at so slow a rate that tens of thousands of years are required to produce appreciable amounts of variation in those latitudes. Near the poles, however, where the effect of change in the earth's position with regard to the sun, from the decreasing convexity of the terrestrial surface consequent upon its oblato-spheroidal figure, must obviously have considerable influence upon climate, and where, too, the slighest fluctuations of temperature must be perceptible as affecting in various ways the relations of the inhabitants, the comparatively recent migration of Esquimaux tribes from places much further north to the localities they now inhabit, and the impending necessity of a still further retreat southward as the only means of averting prospective extermination through the increasing rigor of climate, show that a climatic change is in progress there which cannot consistently be referred to any other known agency.

That the transverse rotation serves to produce a sensible effect upon polar climate within the space of a century or two, while a correlative change in lower latitudes is to be detected only in periods of tens of thousands of years at least, is satisfactorily accounted for when we consider the extreme disproportion that exists between the extent of the frigid zones as compared with that of the remainder of the earth's surface. The annual amount of heat received by the earth from the sun is constant, and it follows, therefore, that whatever loss of solar heat one portion of the earth may at any time suffer, some other part must gain. But the quantity lost by polar areas within such limited periods is so small, that although its loss may be sensibly felt in those areas from century to century, when that quantity comes to be disseminated over the vastly more extensive torrid and temperate zones, it becomes entirely inappreciable.

For the same reason, in the times immediately preceding the historical, no indications are to be found in middle latitudes, of any deviation from the order of things now prevailing, either in the organic or inorganic worlds; and it is difficult to locate in time the precise point at which the evidences of climatal change become obvious. But no geologist will be likely to question either the long duration of the several subdivisions of what is called the Post-Tertiary age, or the fact that after they do commence, the traces of a more and more rigorous cold constantly increase as we recede further and further into the past. The age of Bronze and the later portion of the age of Stone, or Neolithic age, show no signs of a climate colder than that of the present time; and, according to Lyell, the first indication we have of the prevalence of a lower temperature in middle latitudes is found in the range of animals fitted like the reindeer for a northern climate, into latitudes further south than those to which they are now restricted.

In the intermediate portion of the age of Stone, the reindeer and other northern species, while showing no diminution of number in what now seems their appropriate home, extended their range in Europe to the foot of the Pyrenees.

That this was not a migratory movement necessitated by an absolute decrease in mean temperature, but only a mere overrunning by these species of their present southerly bounds in consequence of the extension of the, to them, more favorable conditions of life afforded by the polar-like winters of increasing obliquity, is evident from the fact that at about the same time the mollusks of the Norwegian seacoast ranged several degrees further to the northward than they now do. An absolute decrease in the terrestrial temperature must inevitably have tended towards a general removal of all organic forms southward; while, on the contrary, the more equal distribution of the solar influence over the earth, together with the colder winters and warmer summers of greater obliquity, would certainly tend to the diffusion of many species over a much larger territory than they now occupy. The increased rigor of winter would carry northern types southward, especially upon the land, while the heats of summer would tend to increase the range northward of southern forms — an effect more likely to be observable among marine tribes, there being, as obliquity increased, less and less difference between the temperature of equatorial and polar oceans than there now is.

This era appears also to be characterized by the extinction of many species of large vegetable-feeding mammiferous animals, such, for instance, as the Mastodon of the northern hemisphere, the Megatherium of South America, and the Diprotodon of Australia. The sharp contrasts of seasons, as compared with those of our own time, appear to have been so modified and mitigated from the still more marked yearly vicissitudes of palæolithic and glacial times, as to accomplish their extermination, either by acting directly on the animals themselves, or more indirectly on the plants which furnished their subsistence, they being so closely adapted in structure and habit to the peculiar climatal conditions of the Glacial period, that a gradual change therefrom inevitably entailed their gradual destruction. Compared with their nearest living analogues, they were, for the most part, essentially tropical animals, and clearly must have required a hot climate for their development, such climate being no less essential to the growth of the plants which furnished sustenance to their huge bodies.

But the shaggy covering of hair, fur, and wool, with which they were provided, so compounded as to afford the best possible protection against extreme cold, is sufficient evidence that they were obliged to endure such extreme of cold; for, in the light of analogies universally obtaining throughout the whole animal kingdom, it is highly absurd to suppose that nature, always chary of useless effort, would provide these creatures with so complex and heavy a protection against cold in the absence of cold. The food of these animals, to the nature of which we fortunately have a clew, indicates their ability to subsist, for a part of the year at least, on the ligneous tissues of trees and shrubs; and the peculiar conditions in which their remains have been preserved and handed down,—some of them before becoming enveloped in their icy cerements, having evidently been subjected to the influence of a remarkably hot and drying atmosphere; others, from partial decomposition having taken place before such envelopment, exhibiting the effects of a warm and moist condition of the air; and still others, which, from their entire and recent or fresh appearance, must have been, on their demise, immediately immured in their frozen covering,—all furnish important collateral testimony to the prevalence of an extreme diversity of seasons.

It is by no means intended to restrict the destruction of these tribes of animals to the period under consideration. Indeed, it is very probable that the process of extinction may have commenced as far back even as the beginning of the Palæolithic era, and its complete consummation have been attained only in the ages immediately preceding the historical.

The changes in the aspect of the earth's surface between the intermediate portion of the age of Stone or the Reindeer period and the present time, are, as we would naturally suppose, comparatively slight.

In the earlier Stone or Palæolithic age, when, according to our view, the earth's axis must have been inclined at an angle of about 45° to the plane of the ecliptic, we find ourselves among the conditions peculiar to the Glacial period. The succession of seasons of extraordinary heat and extraordinary cold, arising from such a large inclination, must have had the effect to produce comparatively large annual floods, which, although on a far less grand scale than those of the Glacial period, are gradually completing the transpositions of that great epoch.

The fallacy of the human judgment can, perhaps, nowhere be more strikingly illustrated than in the fact that, as we contemplate the results which must follow this and still larger degrees of inclination, our first impulse is to exclaim that such results are irreconcilably inconsistent with the continuance of life; for it is easy to imagine that if such climatal conditions were now to be suddenly precipitated upon the earth, very few, if any, of the species of plants and animals inhabiting it could possibly survive. The difficulty is, that we involuntarily assume the present status as a fixed standard, to which all other eras must conform, when in reality it is but a passing view in the ever-changing panorama of nature.

The position is clearly untenable that the organisms existing previous to the Glacial epoch, or even those of our own time, could not, by means of extremely slow and uniform climatal changes tending in that direction, become so modified as to endure, or rather to exhibit, a close co-adaptation to its peculiar conditions; for we know that those conditions, so far from being inimical to organic development, for some reason not yet clearly seen, were unusually favorable to such development. Throughout the whole duration of the Ice period flourished in abundance, over the whole earth, the most bulky species of the higher types of animal life. The extinct species of the genera elephant, rhinoceros, hippopotamus, &c., in the northern hemisphere, comprising the European, Asiatic, North African, and North American provinces; the Megatherium, Megalonyx, Glyptodon, Mylodon, Toxodon, and Macrauchenia of South America; and the Diprotodon and Nototherium of Australia, all contemporaneous forms, referable to Post-Glacial, Glacial, and Upper Miocene times, were evidently the specific products of the peculiar climatal conditions then prevailing. The simultaneous development of these large animals, including species peculiar to the several divisions of the earth's surface, all attaining great bulk, flourishing throughout a vast epoch, and finally becoming extinct at the same time, points emphatically to the operation of a common cause. However widely separated, however specifically differing from each other, these vegetable-feeding tribes have all been affected alike over the whole extent of the earth. No cause can be imagined so likely to give size to these animals as an abundant supply of appropriate food; and who will venture to affirm that the climatal conditions resulting from great obliquity and the more equal distribution of the solar influence over the earth might not develop a flora in every region such as would furnish the needed supply? We have no direct means of ascertaining what the effect of a continuous day of from three to six months' duration may have been upon the vegetable kingdom; but if we regard night as likely to retard growth, a period exempt from its effects, with abundant moisture, — those times exhibiting no signs of deficiency in this respect, — can very reasonably be conceived as highly favorable to vegetable development. The severe but short winters might, indeed, possibly tend to prevent plants from attaining large size; but the seasons of rest afforded by them may have had the effect to stimulate development during the comparatively long season of growth, while the excess of solar influence during the growing season, for aught we know, may possibly have had a material influence in increasing the nutritive properties of vegetable products.

The moment we attempt to account for these facts under the hypothesis that the appearances indicative of former vicissitudes of climate result from absolute variations in temperature, the earth retaining its present position with regard to the sun, and from unknown extrinsic causes experiencing vast alternate epochs of heat and of cold, affecting relatively all parts of the earth's surface, we become involved in the most perplexing and inexplicable difficulties. If the Glacial period was one of severe unmitigated cold, how could these animals have maintained existence anywhere on the earth's surface? and, as an even more perplexing enigma, how could they have extended their range almost, if not quite, to the poles? If they could by any means have been enabled to endure the intense cold, whence the plants to afford them sustenance? The difficulty becomes none the less by supposing the cold to have been but little in excess of that of the present time. If the heat of the sun were now to gradually decrease from year to year, no one could possibly doubt that the effect would be to determine a universal migration of the whole organic world southward, and a corresponding increase of temperature would induce a similar movement northward. It is needless to observe that no such general migratory movements have ever occurred in connection with the coming on and going off of the Glacial period. On the contrary, during its most excessive rigor, modified types of tropical herbivorous animals inhabited all parts of the earth, and, of course, must everywhere have found their appropriate food.

How rational, how natural all these phenomena appear, viewed in the light afforded by the theory of the transverse rotation of the earth! On the other hand, what difficulties, what perplexities are encountered in the endeavor to reduce them to system, either by means of those limited secondary local causes of variation in climate, which at best can only affect limited areas of the earth's surface at any one period, and only for limited amounts of time, or by recourse to the theory of absolute variations in its temperature! He who, in the past history of the earth, endeavors to account for each and every separate departure from the standard of to-day by a separate and distinct theory — who shall, for instance, seek to account for a profuse vegetation at the poles, ice-action at the equator, or apparent abnormal warmth or frigidity in middle latitudes, and other like departures from the prevailing status, each by a separate concatenation of minor agencies, — has, indeed, a task before him to which the fabled labors of the Danaides, or of Sisyphus, can hardly be compared.

The opinion that the Palæolithic and subsequent ages were characterized by great extremes of heat and cold, receives a remarkable confirmation in the fact of the trogloditic tendencies of the races of men by whom the world was then peopled. Either in the natural or in the civilized state, men are not prone to take up their abode beneath the surface of the earth. The labor involved in the construction of underground habitations, together with their evident discomfort, would, in the absence of pressing necessity, deter men, and especially savage men, always averse to hard labor, from a general recourse to this kind of residence. We can hardly conceive of climatic conditions under the present status, the result of absolute variations of temperature, affecting alike, relatively, all latitudes, that would lead men to take up their abode beneath the surface of the ground.

There seems to be no doubt, however, that the early races were troglodites. Caves and artificial subterranean excavations, which once, evidently, served as human habitations, abound in the eastern hemisphere. India, Afghanistan, and Syria, in Asia; England, France, Germany, and Norway, in Europe; are known to have been once inhabited by dwellers in caves, and antiquarians are familiar with the traces of their abodes; while it would appear that Egypt derived her peculiar order of architecture from the idea of excavations in the earth.

The most rational explanation of this general recourse to under-ground residences, is to view them as places of retreat from the great extremes of heat and cold characteristic of a highly variable climate, affording as they would a refuge from the more than tropical heats of midsummer, as well as from the excessive cold of the polar-like winters. The trogloditic propensities of these early progenitors of the human race is alone explicable on this theory; and the evolutionist may, perhaps, find it worth while to speculate how far this forced periodical congregation of the members of the human family, in close contact in narrow quarters, may have contributed to its intellectual advancement.

The amount of significance accorded to those memorials of the so-called Post-Pliocene Glacial period which have been transmitted to our times, as indicative of its climatal state, varies with the conflicting views entertained regarding it. There are those who believe that they point to nothing less than a degree of frigidity attainable only through the almost entire absence, or neutralization, of the solar influence. No rational or in any degree satisfactory hypothesis has been, nor, we may add, can be advanced to account for such a withdrawal of the solar light and heat from the earth; but, pointing to the enduring and obvious traces, engraved upon the flinty rocks, of the severe cold of the winters of those times, and ignoring the many marks indicating the warmth of the summers, less pronounced and of a more evanescent nature as they are, they contend that nothing but an intense degree of cold, affecting in relative proportion all latitudes, could have produced the monuments in question. This view, which seeks to exaggerate the phenomena of the Glacial period into evidence of an era of unmitigated cold of long duration, or a "geologic winter," is opposed to the fundamental laws controlling the material universe, and also to a just rationale of the attendant correlative facts.

In the first place, such a theory involves a very large degree of solar variability, for by no other natural means could such a winter have been brought about.

The idea of the sun being a changeable body, subject to irregular periodical mutation, is repugnant to our innate conviction of the fitness of things, and is wholly unsupported by any species of evidence whatever.

If it is possible for any proposition, unsupported by collateral evidence to challenge our assent, and to which such assent cannot be construed as indicating weakness or credulity, that proportion is the one which asserts the invariability of the solar influence.

There is no necessity, however, of accepting it as a matter of faith. The best of evidence is adducible in its support. The fact that the sun has continued through decades, if not hundreds of millions of years to shed upon the earth its benign influences in the precise adequate proportion essential to the needs of the organisms inhabiting it, is a sufficient attestation of its immutable and constant nature.

But even were we to waive this consideration, and admit the possibility of solar variation, the problem, far from being simplified and rendered comformable to the facts, on the contrary, becomes still more involved and perplexing, and still further removed from a rational and satisfactory solution. If the climatal phenomena of the Glacial period were determined by an absolute decrease of the terrestrial temperature consequent on a decrease of solar heat, how could a vegetation so abundant and varied as to suggest the idea of a tropical climate, have flourished in high arctic latitudes, and how could numerous species of animals, many of them now restricted to tropical and sub-tropical regions, have extended their range to the poles during its greater rigor? and why did the gradual restoration of the solar influence to its assumed normal value have the effect to exterminate or force a general migration southward of this polar fauna and flora of the Glacial period? If the earth and sun at that time held their present relative positions, the heat of the sun at the equator being to that at the poles as one hundred and fifteen to fourteen, when the temperature became sufficiently high at the poles to support such a fauna and flora, the heat at the equator must have increased in the same ratio. Where are the indications of such a total extermination, such a complete incineration of all organic structures as must have followed the high degree of heat that must have prevailed not only at the equator, but which must also have been quite intense enough to have accomplished that end throughout nearly, if not the whole, of the temperate zones? The fact is, however, that so far from this extreme heat having prevailed at the equator, the climate there, at that time, exhibited a degree of rigor adequate to the formation of a gigantic glacial system. The problem, then, for the advocates of the cosmic winter theory, stripped of verbiage, and reduced to a simple practical statement, is to show how the sun, following approximately his present path in the heavens, if subject to change, could have so varied as to produce synchronously a cosmic winter over the whole earth, the tropics not being exempt from its rigors, a climate at the poles so nearly resembling the tropical as to be for a long time considered as such, and at the same time an analogy or general resemblance of climate from the equator to the poles. There is no possible way to avoid this issue, and the enigma can only be resolved through the subversion of the laws of Nature, and the re-establishment of the reign of Chaos.

The same argument applies with undiminished force to the theories of those who, instead of exaggerating the evidences of former climatal changes, endeavor to belittle and reduce them to their lowest possible terms, in order to account for them through the agency of clearly inadequate secondary and local causes.

In addition to the difficulties encountered by the advocates of the cosmic winter theory, these last, as before suggested, have also the labor of accounting for each and every separate departure from the now prevailing status, —

such departures being considered as anomalous localized deviations from the established order, — by a separate and distinct chain of causes. It is not claimed that any of these alone are of adequate importance to produce the changes in question; but it is held that in some inexplicable manner a number of these minor influences may, by their united concurrent action tending to that end, at one time have brought about seasons of excessive cold, and at another, by their combined action, tending in the opposite direction, have had the effect to produce seasons or eras of unusual warmth.

These assumed causes of the climatal vicissitudes of former times are, possible changes in the relative proportions of land and sea and in the elevation of the land, variations in the degree of eccentricity of the earth's orbit, the precession of the equinoxes, and an assumed limited amount of variation in the position of the earth's axis of rotation. The question now is, are the phenomena of the Glacial period conformable to such an hypothesis?

Of the first-mentioned, or geographical causes of climatal change, to which Sir C. Lyell has accorded a controlling influence, it may be said, in the first place, that the existing general arrangement of land and water is extremely ancient. Notwithstanding the more or less localized oscillations of level that have been, and are still, in constant progress, "the principal masses of land," in the words of the above-mentioned author, "have continued so long above water, that each of them is now tenanted by a distinct set of animals and plants." It may safely be assumed, then, that from the commencement of the Pliocene period to the present time, the changes that have occurred either in the outlines of the great bodies of land, or their mean elevation above the level of the sea, have been very slight, compared with the amount necessary to sensibly affect climate. Whatever changes of this character may have occurred within this time have been restricted to limited areas and comparatively short periods, while the climatal vicissitudes of the Glacial epoch were general, affecting all parts of the globe at the same time.

We may imagine, although we can by no means be certain of the fact, that if a large proportion of the land of the northern hemisphere should sink beneath the waves of the ocean, and an upheaval of corresponding extent should at the same time take place antipodal thereto, the same might tend to equalize the climatic difference now existing between those two divisions of the earth's surface. The effects of such an equalization of temperature, however, would exhibit neither resemblance nor analogy to the phenomena of the Glacial period. Nay, more: were we to conceive of all the land of the globe, with the exception, perhaps, of a single medium-sized island at one of the poles, concentrated in a broad belt about its equatorial circumference, or, on the other hand, excepting a like island at the equator, divided into two equal portions, and each forming a circular continent around one of the poles, in neither case would we be justified in the expectation of finding upon the island phenomena analogous to those of the Glacial period, assuming the earth as continuing to hold its present relative position with regard to the sun. In the first instance, the climate of the island at the pole might exhibit a small departure from the mean temperature of the same latitude under the present geographical status; but it would, unquestionably, be the height of absurdity to suppose that either the polar island or the polar continents would, in consequence of such geographical arrangements, experience a climate similar to that now enjoyed within the tropics; and no less groundless and absurd would be the assumption that, as a consequence of their positions, a climate would ensue, either in the equatorial continent or equatorial island, such as should admit of the formation of glaciers at ordinary levels, or, at least, at a slight elevation above the same, under the direct influence of a vertical sun.

If these greatest possible extremes of geographical change can only be attended with comparatively trivial climatal effects, the conclusion is certainly justified that vastly less amounts of such change are wholly inadequate to cause any appreciable variation whatever in terrestrial climate.

That the climatal peculiarities of the Glacial period were not dependent on geographical mutation is also shown by the fact that the difference in the mean temperature of the same latitudes of Europe and of Eastern North America continued to maintain throughout the whole of that era the same proportion it did both before and after the same;

the inference being legitimate (this difference being due to causes of a geographical nature), that if the Glacial period was the result of the operation of the same class of agents causing the contrast in question, acting with the vastly increased force required in the premises, the relations on which such difference depends must have been more or less disturbed, if not completely annihilated.

It is well known that the difference in the climates of Europe and North America is due, on the one hand, to the fact that Europe is separated from circumpolar lands by the waters of the Arctic Ocean, and is the recipient of a large amount of heat transported from the tropics by the Gulf Stream; and on the other, by the extension of the land of the North American continent into the arctic circle, reaching with interrupted continuity as far as man has as yet succeeded in penetrating in the direction of the pole, and also by the refrigerating effects of a polar oceanic current traversing its eastern shore on its way towards the tropics. The direction of the above-named marine currents is determined in a large measure by the shape of the contiguous continents. Now, if this specific difference between the climates of Europe and of North America has remained the same from an era anterior to and throughout the whole duration of the Glacial period, down to the present time, both the Gulf Stream and its counter arctic current must, through the same period, have continued to flow in their present channels. The outline of at least the Atlantic shore of the western continent cannot have materially varied, and the same may be said in regard to the extension towards the north of both continents. Now, can it be reasonably supposed that an amount of geographical change sufficiently large to constitute it the controlling cause of the phenomena of the Glacial period could have taken place in the northern hemisphere, without more or less disturbance of the complicated arrangements on which that difference depends? Admitting this exceedingly improbable supposition, the partial and questionable solution it affords covers only half the ground, — the northern hemisphere, — and leaves us still under the necessity of framing a new and separate hypothesis to account for its contemporaneous occurrence in the southern half of the globe.

The proposition always holding good that, under the present status, circumpolar areas must exhibit a frigid, those of the equator a torrid, and those of middle latitudes a temperate climate, whatever may chance to be the physical conformation of the terrestrial surface, what can be plainer than that the Glacial period, so far from being wholly or in part due to such a limited and strictly localized agency, the possible effects of which can only have been, at best, comparatively small, must have been the result of a grand primary cause, affecting, relatively, all parts of the earth alike, and acting independently of and superior to this and other like minor influences? If causes and their effects are commensurable, where the latter are general and unlimited they cannot be consequent on limited and localized causation.

The agency assumed to be next in importance to that just considered, in determining the phenomena of the Glacial period, is that founded on changes in the distance between the sun and earth resulting from variation in the form of the terrestrial orbit. Being elliptical, the sun occupying one of the foci of the ellipse, the earth, at one point in its annual course around the sun, must be nearer to it than when at the opposite point, supposing the two to constitute the extremities of its major axis. The difference between the greatest and least distances, or eccentricity of orbit, amounting at the present time to about three millions of miles, is not constant, but is now gradually diminishing at a slow and irregular rate, and will continue to decrease for about two hundred and forty centuries, attaining in that time its minimum value, about five hundred thousand miles, the orbit becoming then as nearly circular in form as it ever can. On the other hand, an increase is possible in the opposite direction to a maximum of fourteen millions of miles. The movement is not uniform from one extreme to the other, but proceeds in a highly irregular manner, sometimes traversing nearly the whole extent of range in a period of five hundred centuries, and at another being for half a million of years restricted within comparatively narrow limits, the least eccentricity in that time being three millions, and the highest seven and three-quarter millions of miles. These perturbations are due to the attraction of the nearer and larger planets, and the effect is, that the distance of the earth from the sun, while always preserving an invariable mean, is, within the annual limit, constantly changing, and may vary from one half million to fourteen millions of miles.

Theoretically, the intensity of the solar influence diminishes as the square of the distance increases, but practically this does not seem to hold good; for it is asserted that the whole earth is warmer in June, when furthest from the sun, than it is in December, when nearest — a fact forcibly illustrating how little influence variations in eccentricity can have exerted in producing the climatic vicissitudes of the Glacial period. If the hypothetical effect of an eccentricity of three millions of miles is now found to be more than neutralized by other agencies, certainly no one has the right to insist that larger possible amounts were, to any great extent, instrumental in producing the wonderful phenomena of that era.

The most convincing evidence of the erroneous nature of the theory which accepts variation in eccentricity as a concurrent agent in producing former climatic vicissitudes is found when we compare the probable date and duration of the Glacial period with the dates and duration of past variations in eccentricity. Geologists, basing their conjectures upon a general survey of their peculiar field of research, have become accustomed to speak of the Glacial period as dating back a hundred or two thousands of years, and as covering in duration a period of time amounting to several thousands of centuries. Indefinite as these figures are, they show, under the circumstances, a remarkable approximation to those arrived at by means of our system of computation, which places its termination one hundred and sixty-one thousand five hundred years ago, and assigns to it a duration of six thousand seven hundred and fifty centuries; its commencement, therefore, being about seven hundred and thirty-six thousand five hundred years ago.

By turning again to the table on page 176, showing the variations in eccentricity that have taken place during the last million years, and going back a few thousand years previous to the commencement of the Glacial period, we find the eccentricity of the earth's orbit to be ten and one-half millions of miles; having, within a comparatively brief interval, undergone a series of extreme fluctuations, suddenly becoming reduced to four millions, or a little more than its present value, and for five thousand centuries, or nearly the whole duration of the Glacial period, varying but little from that as a mean, and at its close exhibiting, for about one hundred centuries, a large amount, — about ten and a half millions, — retaining a high value down to nearly the present time.

From the more permanent character of the traces of cold than those of warmth, the geological record indicates an apparent gradual and uniform decrease of the temperature of the globe from the close of the Miocene period to the intermediate point of the Glacial epoch, and a gradual and uniform increase of the same from that point to near the present time. In order, therefore, to show that eccentricity has in any degree been responsible for these changes, the table should indicate, from the commencement of the million years to the initial point of the Glacial period, a uniformly low degree of eccentricity; it should then gradually increase, reaching its highest value at the middle of that period, about five hundred thousand years ago; thence gradually decreasing to within about one hundred and sixty thousand years of the present time, reaching then a low amount, and remaining uniform from that time to this.

Nothing, then, can be more hopeless than the attempt to establish a synchronism between an astronomical perturbation so irregular and capricious in its nature as the one under consideration, and the slow and uniform changes from certain climatic states to those of a decidedly opposite character that have diversified the former history of the earth. Even could an accidental correspondence be shown, the grave doubts that are justifiable in relation to its efficiency as an agent of climatic change must constitute it a most unreliable element in the solution of the grand problem.

Another astronomical motion, which, it is supposed, may have exerted some influence in bringing about the climatic changes of former times, is the precession of the equinoxes. "Its effect," says Lyell, "is to cause the different seasons of the northern and southern hemisphere to coincide successively with all the points through which the earth passes in its orbit round the sun," determining, in stated portions of time, reversals of the relative climatic state of the two hemispheres. On the supposition that the so-called Post-Pliocene Glacial period was an exceptional departure from the normal standard, brought about by a fortuitous concurrence of a certain number of minor agencies, it is assumed that it may have occurred in the northern hemisphere, when the winters there were in aphelion, contemporaneously with an extreme eccentricity of the terrestrial orbit.

Obviously, a most conclusive objection to the above view is to be found in the fact that the peculiar conditions of the Glacial period pervaded the whole earth at one and the same time, instead of being limited to any portion of it. There can be no doubt that when the glaciations and denudations of the Ice period were progressing in the northern hemisphere, the same processes, on a corresponding scale of magnitude, were going on in the southern. So, also, when the climatal conditions of that time had the effect in Europe, Asia, and North America, to evolve the Mastodon and his bulky contemporaries, from pre-existent forms, they at the same time, and in the same manner, produced Megatherium, Diprotodon, and their compeers in South America and Australia; and when the conditions on which the existence of these animals seemed to depend became inoperative or so modified as to entail their extinction in one place, the same cause brought about their extermination all over the earth. Cause and effect are commensurable. The effects in this case were general, and their determining cause can have been no less. No hypothesis, therefore, is adequate to account for the phenomena of the Glacial period that fails to meet this requirement.

Another objection, of perhaps equal force, arises from the comparatively short periods of time comprehended in complete cycles of precession. Were this movement to proceed independently, it would require, for its successive accomplishment, periods of twenty-five thousand eight hundred and sixty-eight years; but being, as it is, complicated with what Sir J. Herschel terms the movement of aphelion, or revolution of the apsides,—a motion effecting a gradual displacement of the major axis of the earth's orbit, which operates to shorten the time,—it is completed in intervals of twenty-one thousand years. Thirty-two of these cycles of precession would, therefore, be included within the probable duration of the Glacial period; and as the supposed effect could not act concurrently with extreme eccentricity for more than one-fourth of the whole period, or fifty-five centuries, being in direct opposition to the same for an equal length of time, and neutral the remainder, the effort to show that the precession of the equinoxes could have had value as an auxiliary agent in bringing about the climatal conditions of an epoch covering some six thousand seven hundred and fifty centuries, must inevitably fail of success.

Clearly, then, the phenomena of the Glacial period cannot have been the result either of the separate or conjoint action of these limited and incongruous agencies; and not more amenable to them either, are those hypotheses which depend upon absolute changes in the amount of terrestrial heat, whether the same be the effect of fancied variations in the temperature of space, or of the even more improbable variability of the solar influence. The hypothesis of Mr. Evans, which suggests a change of direction of the axis of a supposititious exterior shell or crust of the earth, by the sliding of the same over an internal fluid nucleus, the axis of diurnal rotation remaining unchanged, were the event possible, is perhaps the only one heretofore advanced under the assumption of the stability of the inclination of the terrestrial axis of rotation that can be made to assume even the slightest air of probability. The oblato-spheroidal form of the earth, however, and the demonstrated solidity of by far the larger part of its mass, aside from the inherent improbability of the proposition, attests, with a sufficient degree of certainty, the utter impossibility of such a movement of the exterior portion of the material of the earth.

While, therefore, fatal and insurmountable objections oppose themselves to all of the methods heretofore employed to account for former diversity of climate, the complete reduction of all the attendant phenomena connected with them to a consistent and rational system is easily and naturally effected through the theory of the transverse rotation of the earth. Viewing those changes as effects of such a motion, every real difficulty vanishes, and those suggested by the imagination are readily resolved in the light of modern science.

It is true that authority, although with unsupported

dictum, has pronounced such a movement impossible ; and as we look abroad over the face of nature, and behold the close co-adaptation subsisting between all the diversified forms of life and their surrounding conditions, the objection which evidently inspired that dictum forcibly presents itself to our minds.

There is reason to believe — indeed, it may be considered quite certain — that a sudden change from the present status to that which would obtain under an inclination of 45° or more, would entail an extensive destruction of the living inhabitants of the earth. But this fact does not authorize the inference that the earth's axis has never assumed such a position, because the consequences of sudden transitions have alone been taken into account, while the possible effect of an almost inconceivably slow and uniform movement from one position to the other has never before received attention. Thus it is that while our view is limited to the present time, and to our immediate surroundings, the transverse revolution of the earth seems an utter impossibility ; but a wider scope, however, discloses not only its possibility, but also how absolutely essential it is, in the terrestrial economy, as a primary element in determining the gradual evolution of higher and higher types of life out of pre-existing inferior forms — a process coeval with the inception of life, as far as we know, and, extending throughout all subsequent ages, down to our own time.

Nothing can be more reasonable than to suppose that if the earth has always held its present position with regard to the sun, and, as a consequence, each and every latitude has exhibited through all time an unvarying climate, the organisms on any given parallel of latitude, after having reached the point of closest adaptation to the prevailing conditions (allowing those organisms not to have been created in perfect harmony with them), would, thereafter, indefinitely retain the structure and habits best fitting them for that particular climate. Even an original innate tendency to vary, must, under such circumstances, grow gradually weaker and weaker, and in process of time become entirely eliminated. In the absence, in the organic world, of a change inducing force, we might reasonably expect to see essentially the same forms inhabiting the same areas, from year to year, and from era to era, throughout whole geological ages.

But such an inert and unprogressive state has at no time characterized the living inhabitants of the earth. From the dawn of life upon the planet, there has been a constant progressive change in the organic world. This continual modification, this uninterrupted substitution of new and improved types in the place of old and inferior forms, is the result of laws whose basis is the ever-changing climatal conditions determined by the transverse rotation of the earth; and it is these changes which in times past have constituted the grand agency through which the flora of one geological period has been transformed into that of its successor. The same force acting directly on the more plastic organisms of the animal kingdom, and also indirectly as affecting the plants furnishing their subsistence, has effected more rapid transitions in that than in the vegetable world. But it is climate, in both instances, which constitutes the primary or antecedent force, producing the changes or chance mutation, while the laws of variation, as laid down by Darwin and others, give progressive direction to them, which else might as often tend to degeneracy as to improvement.

From the apparently fixed and stable character of the material world and its surrounding conditions, it is difficult to divest the mind of the idea that the transitions to and from those eras of past time exhibiting such wide departure from the prevailing status as does the Glacial period, must have been more or less sudden and catastrophal in their nature. We are certain that these epochs of widely diverse conditions have intervened in the past; and, under any hypothesis, we cannot but admit the possibility of their future recurrence. How natural it is, then, to suppose that the present state of things will continue absolutely unchanged for an indefinite period, or until the agent of change, whatever it may be, suddenly assuming an active form, shall inaugurate a new and different status. Thus it is that we so frequently hear such expressions as "the commencement of the Glacial period," "the close" or "breaking up of the geologic winter," &c., suggesting the idea of rapid change. This serious error arises wholly from a misconception of the amount of time involved, for the analogy is perfect between the present annual change from summer to winter, and the change from the geological period of repose, or summer, to the era of transition, or winter. If the time requisite in the former instance may be estimated in weeks, thousands of centuries alone suffice in the latter; and we are to continually bear in mind that at no time do these changes have a greater or more marked effect upon the organic world than at any other, or than they now do. Although the effect is entirely inappreciable to us in our own persons or in our surroundings, yet we are, notwithstanding, just as much the subjects of them, and in exactly the same proportion, as the inhabitants of the earth at any former time have been, or those of any subsequent period of its history will be. If we do not, in our own persons or in our surroundings, perceive any consequences of the transverse rotation of the earth, we have no right to assume them to be more apparent at any other point in the movement.

A broad view, therefore, of the universal capacity of the organic world to adapt itself by modification to the ever-varying conditions of life, and of the vast amount of time allotted by the exceedingly slow rate at which those conditions change to effectuate such modification, shows us, rationally and satisfactorily, how the course of life has continued throughout all the vicissitudes of the Glacial period, with no greater apparent interrupted continuity than during periods which we might naturally suppose to be much more favorable to organic existence.

The amount of modification needed to carry life through the annual extremes of heat and cold, incident to the Glacial period, is very much less than might at first be supposed. Animals furnished with an adequate supply of appropriate food readily accommodate themselves to the most diverse climates; and although, in the case of plants in a state of nature, variation seems to proceed at a much slower rate than in the animal kingdom, proportionally less change would be necessary to adapt them to great excesses of heat and cold. When the vegetation of latitudes where the winter temperature falls below 32° F. has assumed its winter or hibernatory state, the germs of future growth, whether in the form of seeds, spores, or buds, seem to be capable of withstanding any conceivable degree of cold; nor does it matter whether the dormant condition continues for a longer or shorter time, if it be succeeded by a favorable period of growth. Evidently it can make no difference whatever to plants, while in the dormant state, whether alternations of day and night shall transpire each twenty-four hours or not, or whether the temperature remains near the freezing point, or ranges far below it. Indeed, there seems to be no ground to assume that, were polar winters to intervene at the present time between the fortieth and fiftieth parallels of latitude, any great change in the prevailing flora would ensue if the summer temperature were so modified as to be adequate to liquefy whatever excess there might be of snow and ice, as well as to exhibit the ordinary warmth of the usual season of growth. There are no means of determining with precision the effect of continuous day, during a portion or the whole of summer, upon plants; but the wonderful rapidity of vegetable development in the long days of the summers of high latitudes can only be ascribed to the long continuance of the sun above the horizon, and the brief duration of the nights. It has been supposed that night is essential to the growth of plants, as it is, by experiment, observed that they then absorb their supplies of oxygen. If, at the present time, the absorption of this element takes place only at night, it is to be regarded only as a mere habit, for doubtless the process can go on in the absence of the solar influence, no matter what may be the cause of such absence. With the immense evaporation that must be taking place during a day months in duration, it is evident that cloudy weather must prevail for a considerable part of the time; and in our ignorance of the laws of vegetable growth, it would be an unwarranted assumption to conclude that this or any other absorptive or assimilative process connected with the growth of plants may not proceed as favorably under the conditions arising from coincidence of the earth's axis with the ecliptic, as under those now prevailing.

The rationale of the phenomena of superficial inorganic change incident to the Glacial period is much more consonant with the theory of the transverse revolution of the earth than any other. Comparatively slight annual increments of change, aggregating, in a period of six hundred and seventy-five thousand years, to an almost entire reconstruction of the terrestrial surface, is perhaps the only natural way in which such a transformation could have been effected without a more or less complete interruption of the ordinary processes of nature.

All the phenomena of the physical world are the result of determinate natural causes, to the knowledge of which, by a proper use of our intellectual faculties, it is possible for us to attain. The ploughman, unlearned in the physical sciences, whose furrows are obstructed in the midst of an extensive plain by a huge isolated mass of rock, solves the mystery — if the fact should, by chance, assume that form to him — by deciding that the boulder was originally created upon the spot where he finds it. A geologist, viewing the same rock, finds it impossible to accept the ploughman's theory. He searches for, and at a great distance perhaps from the boulder, discovers the parent ledge from which it was originally derived. The question then recurs, By what means has the ponderous block been transported to so great a distance? or, How has it been moved at all? It cannot of itself have rolled or slid for miles, perhaps, along the level plain, to its present position. In his search among the various forces of nature, he finds one, and only one rational method of accounting for its removal. Evidently its change of position was effected through the agency of the buoyant power of ice; and the truth of this theoretical conclusion is confirmed when he comes to observe, in certain localities, precisely analogous processes of rock transportation going on at the present time.

Probably no geologist will deny that the deposition of geological strata has taken place through the operation of law in a manner analogous, in a general sense, to that of the erratic boulder above mentioned. Why, then, is not the determining cause or law as discoverable in one instance as it is in the other? A geological formation is not an accident. It is a series of layers or strata deposed in a regular, determinate manner, from the lowest to the uppermost members, and can be nothing more or less than the result of a corresponding series of climatal vicissitudes, to which the earth has been at some time subjected. These great geological series are all homogeneal. Whatever has caused one of them has caused all of the rest. The effect being identical in each instance, the cause must also be identical. The same round of physical change, recurring over and over again, denotes repetitions of the one producing cause. If one geological group may have been consequent upon climatic conditions incident to a change in direction of the axis of the earth's crust, as suggested by Mr. Evans, another by those occasioned by the passage of the solar system through abnormally warm or cold regions of space, and others by geographical changes acting in conjunction with astronomical causes, as urged by Lyell, or even if they all were the product of the last-named agencies, a corresponding want of conformity would be visible in the results. Obviously, the attendant phenomena would not indicate equal portions of time, while they must, on the other hand, unquestionably show the diverse nature of the producing agents. It is needless to add that no such want of uniformity is discernible in the geological record.

A grand geological group or formation may therefore be defined as the sum of the physical changes effected upon the earth's surface by the climatic variations that must inevitably follow the changes of position of the earth with regard to the sun, during such part of a transverse revolution of the earth as will include a complete cycle of such changes of position; and a geological period expresses the amount of time required for the completion of one such cycle.

The principal reasons why we may regard the phenomena of the Glacial period as resulting from a large inclination, varying from 45° to 90°, or axial coincidence with the ecliptic, are as follows: —

First. The evident correspondence in time between the gradual change from the equable climatic conditions of the close of the Miocene period to the greatest apparent cold of the Glacial era, and the subsequent gradual return of warmth, and those climatal effects that must inevitably follow the changes of position of the earth with regard to the sun under the supposition of the continuity or rotary nature of the motion known as the diminution of the obliquity of the ecliptic.

Second. The equal distribution of the solar heat over the whole earth, rendering the polar regions inhabitable, at the same time that a diminished proportion at and near the equator had the effect to produce there glacial phenomena now limited to high latitudes.

Third. The superficial terrestrial changes incident to that time, which could be the result only of the excessive action of ordinary meteorological agencies, dependent on sharp contrasts of summer and winter.

Fourth. The phenomena of the "Inter-Glacial period," the same being confined to middle latitudes, and showing the effects of summers of nine, and winters of three months' duration.

Fifth. Because the uninterrupted continuance of life, intercalated periods of warmth, and, in a word, all the attendant phenomena of that time, are entirely inconsistent with the assumption of an absolute decrease of the terrestrial temperature, but are consistently, naturally, and rationally accounted for on the theory of an invariable mean annual temperature of the whole earth, the annual aggregate of heat being subject to a large degree of variation in the manner of its distribution over the surface of the globe.

In the light of the new system the Pliocene period is not yet completed, but, embracing within its limits the whole of Post-Tertiary times, it will, in the future, extend over such interval as may be required to bring the polar axis of the earth to an inclination of 45° on the other side of the point of perpendicularity with the orbital plane, which point it is now approaching. The whole amount of time, therefore, required to complete the Pliocene period, making no allowance for the possible retardation that may ensue from polar accumulations of ice and snow during the period of equatorial coincidence with the ecliptic, is five hundred and thirteen thousand five hundred years — a vast era, which, with the exception of the effects produced by oscillations of level in the terrestrial surface, must in a large measure remain a blank in the great volume of the earth's history. At its termination, the changes of another glacial epoch will inaugurate a new geological period, and then will commence a new chapter in the mighty work.

It will be in vain to expect geological periods anterior to our own to exhibit evidences of the transverse rotation of the earth as complete in detail as that does. The unquestionable efficiency of each glacial period in eradicating the traces of its predecessors must render futile all attempts to discover the uniform gradations from one climatic state or condition to another; and we cannot reasonably expect more than to be able to trace, with more or less facility, the prevalence of those distinct states, without being able to establish the boundaries between them. But if the assertion is justified that the great geological periods are the direct effect of regularly recurring cycles of events determined by uniform rotations of the earth transverse to its diurnal revolutions, we certainly have the right to expect them to exhibit, as far as possible, a general analogy one with another, as well as similar phenomena. This expectation is, to a remarkable extent, met and satisfied.

The undeviating regularity in the order of succession, observable not only in the great geological groups, but also in the component members of each, indicates with absolute certainty the intervention of a common identical cause. The superimposition of group after group of like series of strata cannot, under any circumstances, have been the product of dissimilar cycles of similar events, or of similar cycles of dissimilar events. The chain of causation that has deposited in their order the strata of one geological period, has in the same manner and by the same means deposited that of all the others; and if those of the Pliocene or present geological era, differing from previous ones (if there is a difference) only in incompleteness, is the result of the transverse revolution of the earth, every other geological period must also be the product of the same cause. As this general uniformity of sequence points with unerring certainty to a common origin, so the testimony furnished by the monument of each great epoch tends, strongly corroborative, to the same end.

The unequivocal character of evidence derived from whatever signs of ice-action incident to any geological period may have escaped the grinding and destructive action of subsequent glacial eras, can hardly be called in question. Erratic boulders, striated and polished rock-surfaces, and other like phenomena, indicate, wherever found, the former presence of a certain agency acting with a certain degree of force. Glacial striations on the rocks of the Old Red Sandstone as surely attest the presence of glaciers and the prevalence of a temperature at which they may be formed, as do like traces at any later period. This standard of comparison is invariable. It is not so, however, with the evidence resting on comparisons instituted between the living organisms of the present time and those that existed millions of years ago, known to us, in the case of animals, only by a greater or less resemblance of the osseous portion of their structure to modern types, and in the vegetable world, in the shape of the leaves and fruit, and the texture of the ligneous portion of plants. It is well known that animals may bear a close osteological resemblance to each other, and yet in external appearance and habit be almost totally unlike; and also that subdivisions of even the same species, both of animals and plants, may become adapted to almost every variety of climate. Still, it cannot be denied that a general view of the fauna and flora of any particular era furnishes us with a tolerably sure guide to its prevailing climatal condition. But we must bear in mind that this standard of comparison is not infallible, and therefore is to be used with caution. Thus, when geologists who believe in absolute changes in the terrestrial temperature assert from purely organic evidence that the mean annual temperature of a certain latitude, in some era of the past, was ten or twenty degrees higher or lower than now, we may, undoubtedly, rather consider that evidence as indicative, on the one hand, of an excess of summer heat, or, on the other, of an extreme winter's cold.

A warm climate has been ascribed by geologists to the times of the Miocene formation. This supposed abnormal excess of warmth may have proceeded from the cause above mentioned, namely, the hot summers produced by a large inclination, the effect of which would be more observable in high latitudes; or, in middle and low latitudes, from the gradual concentration of the heat of the sun in the direction of the equator as the earth's axis of rotation approached perpendicularity to the ecliptic. The countries affected by the latter cause would not only experience an absolute increase of temperature, but the apparent effect of this increase would be heightened by the uniform distribution of the whole annual amount of heat over the whole year; a less quantity being required to produce the same effects under such uniformity of distribution.

The fossil plants of the Upper Miocene formation of Central Europe, according to Lyell, indicate a climate the nearest approach to which, in the present state of the globe, is that experienced in the island of Madeira. Now, the present uniform climate of that island is precisely analogous to what we would expect in middle latitudes under the conditions incident to equatorial coincidence with the ecliptic. Lower Miocene strata exhibit the same appearances of warmth, and probably from the same cause. In all the ages of the earth anterior to the present or Pliocene age, for obvious reasons, the indications of warmth must be more abundant and legible than those of cold, and considerable difficulty may be expected in distinguishing between the effects of a moderate but uniform degree of heat and those of very hot summers alternating with winters of more or less severity.

In the later Miocene times, when Middle Europe was enjoying a warmer, or, rather, a more equable climate than now, places situated nearer to the equator experienced a still greater heat. The fossils of the Siwâlik hills show a prodigious abundance and variety of fossil mammalia, which, with the associated reptiles, according to Lyell, bear witness to a high temperature. That these formations were deposited contemporaneously, and under coincidence of the equator and ecliptic, there can be but little doubt. In the West Indies, however, in a latitude somewhat lower than that of the Siwâlik hills, the fossils of deposits classified as Upper Miocene lead irresistibly to the opinion that there was a much greater analogy in those ages than there is now between the temperature of the West Indies in lat. 18° N. and that of Europe in lat. 48° N.; and this supposed similarity of climate in places separated by thirty degrees of latitude is ascribed to certain changes from the present geographical status supposed to have prevailed at that time. No attempt is made to show, and it is difficult to conceive, how the submergence of a portion of the Isthmus of Panama could produce a climate analogous to that of Europe in Antigua, San Domingo, and Jamaica. Such a climate there could only result from conditions consequent upon coincidence of the polar axis of the earth with the ecliptic. The West Indian deposits alluded to are not, therefore, the upper but the intermediate members of the Miocene group, and were not synchronally contemporaneous with those of Europe and Hindostan.

The evidences of an intercalated glacial period in Miocene times are unequivocal in character. The glacial interval of any geological period is the period of change, not only in the inorganic, but also in the organic world. While a certain limited amount of change in both must be in progress, even at the intermediate point of the intervals of repose, the more pronounced modifications, the grand aggregate of change, must be the result of the variable climatic conditions of the periods of change, or glacial periods. The interval of repose is that dividing or separating two geological periods, and the interval of change occurs in the middle of each. A closer resemblance, therefore, should subsist between the organisms of the later portion of one geological epoch and the earlier portion of its immediate successor, than between even those of its own upper and lower members. A most remarkable confirmation of this view is found in the fact that while there is comparatively little difference between the fossil forms of the Upper Miocene and of the Lower Pliocene, the difference between those of the Upper and Lower Miocene is much more radical and striking, and this rule will probably be found to prevail throughout the whole geological record.

The range of an abundant and varied flora in Miocene times to nearly the eightieth parallel of north latitude, and, inferentially, to the pole itself, — an absolute impossibility, under the present status, — attests, unmistakably, a distribution of the solar influence at that time wholly different from that now prevailing, and such as only can result from a large inclination of the earth's axis, or its coincidence with the ecliptic. When to these proofs is added the testimony of the erratics of the Miocene formation of the hill of the Superga, the fact of a Miocene ice period of equal importance with that of the Pliocene epoch seems sufficiently well authenticated.

It will be hardly worth while to review in detail the evidence pointing to the transverse rotation of the earth, furnished by those geological epochs which preceded the Miocene. The record of one is substantially the record of all, save that as we recede further and further into the past, the pages become more and more mutilated and fragmentary. Each of these periods, like the Miocene, and for the same reason, seems apparently to have enjoyed a climate somewhat warmer than the present. The Pliocene, Miocene, Cretaceous, Oolitic, Liassic, Triassic, and Carboniferous periods all bear witness, in the range of their semi-tropical types of plants and animals to the poles, to an era within each of axial coincidence with the ecliptic, or glacial period; and doubtless, if the geological record is not too far broken, like traces will yet be discovered in all the others. Besides the Pliocene, the Miocene, Eocene, Cretaceous, Triassic, Permian, and Davonian periods, in the enduring and unmistakable traces of ice action indelibly engraved upon their rocky tablets, more directly, but not more conclusively, perhaps, attest the fact of regularly recurring intervals of glaciation; one of the same constituting an essential, integral portion of each of these great geological epochs. The similarity of these epochs, one to another, seems fully established. The glacial phenomena of the Pliocene period can be traced in the eastern and western, the northern and southern hemispheres of the globe, and the remains of all the great epochs are common to all the earth. The cause that has determined, in its glacial period, a portion of one of these epochs, a single one, or all of them, must be universal,—pervading the whole earth. Now, when to these considerations we add the testimony furnished by Agassiz' observations in tropical South America, which, taken in connection with the proofs of contemporaneous warmth at the poles, shows, beyond cavil, that the last glacial period could have been the effect only of the peculiar position of the earth with regard to the sun, consequent on coincidence of the earth's axis with the ecliptic, it does, indeed, seem as if the whole case might safely be left to rest on the evidence that geology alone has incidentally furnished us.

If a just rationale of geological data points to the existence of a slow rotatory motion of the earth transverse to the axis of its diurnal revolution, so the science of astronomy adds an adequate confirmation to the truth of the theory. Observations extending over a period of at least 3,000 years, and probably twice that lapse of time, show that throughout this period such a movement has been going on, and that it has been constantly tending to reduce the value of the inclination of the earth's axis of diurnal rotation, passing through, in that time, about four-fifths of a degree of circular motion. Under the circumstances it may well be assumed that the motion has been and will be continuous as long as the solar system continues to exhibit its present relations. High astronomical authority, however, has pronounced it to be a mere oscillation, confined to very narrow limits; and an examination of the basis of this opinion became necessary.

We may here observe that the simple fact that the earth has a proper motion of the nature in question, and that its course is opposite to, or away from, what would produce the conditions of the Glacial period, and is approaching what must result in a general uniformity of climate, is all that our system absolutely requires of astronomy. Geology settles affirmatively, beyond a shadow of doubt, the question whether that motion is persistive or not; and the truth of the conclusion can in no degree be affected by either our success or failure in assigning a proper cause. No one would think of objecting to the theory of the diurnal rotation of the earth on the ground that its cause remains unknown, nor, should that cause be discovered, would the knowledge of it render more certain the fact of such rotation. The evidence, although conclusive of itself, would be merely cumulative — an addition to what was amply sufficient without it. Should it be possible to demonstrate that the assigned cause, namely, the attraction of the moon, on the excess of matter at the earth's equator, is inadequate to produce the transverse rotation, and we are driven to speculate on the probability of an original primordial impulse, it is no valid objection to our system; while, on the other hand, if an adequate natural cause can be shown to be in operation, which must inevitably produce the result in question, were that result before a matter of doubt, it must then be considered as established beyond cavil.

The attraction of the moon on the spheroidal figure of the earth affords the most simple and natural explanation of the transverse rotation of the earth. We have here a force continually operating in each lunar revolution, to draw the terrestrial pole first in one direction, and then in its opposite, but always a little further one way than the other; the transverse revolution of the earth being determined by aggregations of these small monthly excesses.

It seems evident that the proximity of the moon to the earth is what constitutes it the controlling influence in this matter, as in the case of the tides, notwithstanding its inferiority of bulk; while the larger but more distant masses of the solar system, by their attractions and counter-attractions, exert the same relative influence, serving to accelerate or retard the force of the lunar impulse in one instance in the same degree as in the other.

In the present state of astronomical science, a complete resolution of all the multiplex forces involved in the determination of this motion may be considered extremely difficult, if not absolutely impossible; and advantage has been taken of this state of uncertainty to reject what we conceive to be the true theory, on the ground of its assumed incompatibility with the continued existence of life upon the earth, and also the promise made to Noah that while the earth continues, day and night, and summer and winter, &c., shall not cease.

The plasticity of constitution, or inherent power of self-modification, common to all organic forms, by which they are enabled to adapt themselves to the most diverse conditions of life, together with the immense amount of time allotted in which to effect these adaptations, indicates the fallacy of the first objection; while the prevalence in various parts of the earth, at the present time, of all the diverse states and conditions incident to the whole globe during a geological period, or semi-transverse revolution, is a sufficient refutation of the latter.

The lunar agency, involving continuity, having been denied, it became necessary to substitute another; and the diminution of the obliquity of the ecliptic, so called, was ascribed to the "joint action of all the planets." As variations in the eccentricity of the earth's orbit are held to be the result of the same cause, a correlation between the two effects was suggested. The idea seemed prevalent that the stability of nature — whatever the phrase may signify — was dependent on the continuance of, approximately, the present inclination, and even the acknowledged oscillation of a degree or two had to be offset by a fancied correlative variation in eccentricity. Unfortunately for this hypothesis, in the case of the earth, the two movements, instead of proceeding in opposition, as they ought to counteract each other's influence, are well known to be moving in conjunction, both decreasing in value at the present time.

The assumed correlation between inclination of the polar axis and eccentricity of orbit and variations in the same, is supposed to hold good not only in the case of the earth, but also in that of all the other planets.

If there is that close relationship between the variations in the two elements which must ensue from both being the result of a common cause, it must be shown in one of two ways; that cause must either operate to increase or diminish both at the same time, as is actually seen in the case of the earth, or, according to the theory of compensation, when the value of one element is on the increase, the other should show a corresponding diminution; and in one or the other of these ways, the relation could not fail to be apparent in astronomical tables of the planetary elements. To accord with the hypothesis, they must show, in the case of each and of every planet, either a high or low degree of both inclination and eccentricity, or, on the other hand, a large inclination must be seen to accompany a small eccentricity, and conversely. It hardly need be affirmed that in neither of these ways is any such correlation indicated by those tables. On the contrary, they prove conclusively that there is no correspondence of any nature whatever between them. Of the two planets which show the largest inclination of axis, one has the most and the other the least eccentric orbit in the solar system. One planet exhibits a very small inclination and a very large eccentricity; in another instance, both are in excess; in another, both show a low value; and where two planets exhibit about the same moderate amount of inclination, the orbit of one of them very nearly approaches the maximum of eclipticity while that of the other approximates closely the circular form.

With every allowance for the imperfection of our mental faculties, common sense seems clearly to teach us that it can make no possible difference in the economy of the solar system, whatever effect it may have on their own relations, how the planetary masses, so nearly spherical as they are in form, shall present their surfaces to the sun. As long as they are globes, what possible difference can it make if their axes are all coincident with, perpendicular, or inclined at various angles to the planes of their several orbits? It can make none whatever, and in either of the three cases, each planet must pursue unaffected its invariable annual path.

The sun, whose influence must in the past have continued invariable from year to year, from century to century, and from geological period to geological period, imparts annually a certain determinate amount of heat to the earth. The mode in which this amount is distributed over its surface determines, primarily, the terrestrial climate. In the absence of any agency operating to effect change in the method of distribution of the solar energy, the climate of the earth generally, and that of any portion of the same, must assume and continually maintain static conditions. As long as the annual amount of solar heat is distributed as at present, each latitude will invariably receive the proportion of that amount it now does, and that proportion will be distributed throughout the seasons of the year in the manner now observed. Certain local agencies may vary these results in a comparatively slight degree; but under no conceivable circumstances is it possible, under the present mode of solar heat distribution, for any radical change to transpire in the climatal conditions prevailing at any point on the earth's surface. No one will deny that it would now be a physical impossibility for natural glacial processes to go on at ordinary levels under the vertical sun of the tropics, or for an abundant vegetation analogous to that of low latitudes, with its attendant animal life, to be developed in the immediate vicinity of the poles. If impossible now, it must have been equally so at every other stage of the earth's history, under the same circumstances. The geological record, however, indicates with certainty regularly recurring intervals characterized by climatal conditions productive of these classes of phenomena. The conclusion is, then, inevitable that there must have been a change of status or a departure from the present mode of distributing the annual amount of solar heat over the face of the earth; and such departure necessarily involves change of position of the earth with regard to the sun.

The only possible way in which such change of position can be effected is by a movement of the earth transverse to the axis of its diurnal rotation. Such a motion, determining in stated periods of time complete revolutions of the planet, would produce the phenomena in question, and, if sufficiently slow, would, through their inherent plasticity of constitution, give time for the living inhabitants of the earth to adapt themselves, by self-modification, to the changing conditions of life. We know that a motion of the earth of this description has been in progress for thousands of years — is going on now before our eyes. Admitting its continuity, and tracing the natural effects of such a motion, all the perplexing, anomalistic facts of the geological record — rendered so by efforts to make them conform to a false, imaginary stability of nature — become naturally and readily resolved, and a great natural system is indicated, through which it seems indeed possible that all, or at least nearly all the phenomena of the physical world may be reduced "to the unity of a single principle."

THE END.

www.ingramcontent.com/pod-product-compliance
Lightning Source LLC
LaVergne TN
LVHW010239110826
845151LV00004B/1341
* 9 7 8 1 4 2 5 5 2 4 4 9 4 *